ANGRY WEATHER

Translation by **Sarah Pybus**

FRIEDERIKE OTTO
WITH BENJAMIN VON BRACKEL

DISCARD

ANGRY WEATHER

Heat Waves, Floods, Storms, and the New Science of Climate Change

DAVID SUZUKI INSTITUTE

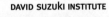

GREYSTONE BOOKS
Vancouver/Berkeley

Greystone Books Ltd.
www.greystonebooks.com

David Suzuki Institute
www.davidsuzukiinstitute.org

Cataloguing data available from Library and Archives Canada
ISBN 978-1-77164-614-7 (cloth)
ISBN 978-1-77164-615-4 (epub)

Copyediting by Dawn Loewen
Proofreading by Jennifer Stewart
Jacket and text design by Nayeli Jimenez
Jacket photograph by Karin Broekhuijsen/Buiten-Beeld/Minden Pictures
Printed and bound in Canada on ancient-forest-friendly paper by Friesens

Greystone Books gratefully acknowledges the Musqueam, Squamish, and Tsleil-Waututh peoples on whose land our office is located.

The translation of this work was supported by a grant from the Goethe-Institut in the framework of the "Books First" program.

Greystone Books thanks the Canada Council for the Arts, the British Columbia Arts Council, the Province of British Columbia through the Book Publishing Tax Credit, and the Government of Canada for supporting our publishing activities.

For my Ottos

CONTENTS

PREFACE

I WROTE THIS BOOK in the spring and early summer of 2018, and revised it during what turned out to be a summer that dramatically changed the conversation we are having in society about climate change. Or to be precise, one we are starting to have on a scale that finally seems to come close to what we need to actually address climate change.

This had nothing to do with me, but the science that is described in my book probably did play a role. During the summer of 2018, people in France, the U.K., Germany, India, North America, and many other places around the world not only experienced what climate change feels like but became aware that the very high temperatures they sought to avoid in the shade were not just weather but part of a changing climate. The following northern hemispheric summer of 2019 again saw heat records being broken throughout Europe, including the more than eighty-year-old U.K. record.

Extreme heat was also a key driver of the bushfires that destroyed lives, livelihoods, and ecosystems in southeast

Australia. As before, new studies undertaken by the team described in this book found that without human-induced climate change, the heat in Australia would have been at least a degree less intense and less than half as likely as in today's climate. Climate change also made the weather conditions leading to the fires overall at least 30 percent more likely, which means that without climate change the devastation these fires wreaked would have been significantly less severe.

This book—which looks at how weather and climate change are linked and how we as scientists can now characterize and quantify humanity's role in extreme events—has become even more relevant and timely than I imagined it to be. Or, in the words of a German radio station, it "provides the arguments for the Fridays for Future movement." Not all of them, certainly, but in this book I describe the birth of a new way of doing climate science. Not only in specialist journals and highly complex reports, but as and when and where people ask scientific questions and need scientific evidence.

Climate change is a fact. We've known this for a very long time, with experiments confirming the greenhouse effect conducted by a largely ignored scientist, Eunice Newton Foote, as early as 1856 and fully quantified by Svante Arrhenius forty years later. We have observed rising global temperatures over the course of the twentieth century, and the science advisory committee of Lyndon Johnson's presidency warned of global warming in 1965.[1] At the very latest, since the 1990s we have been able to attribute these rising global temperatures to greenhouse gases in the atmosphere from the burning of fossil fuels. However, global mean temperature rise is not killing people and ecosystems directly. Thus the one degree of global mean temperature rise we have today is for most of us just a number. It is a powerful and important number, but since we do not experience it directly, this number only allows (and crucially, requires!) us as a global society to tackle climate change with our intellect, not fueled by direct experience and resulting emotions.

Being human, we find that a very hard task at the best of times. It hasn't exactly been the best of times, though, with powerful interests and a lot of money devoted to characterizing the laws of physics as a hoax. Published research from historians shows that leaders in the oil industry knew about the consequences of continuing their business model (digging up fossil fuels to be burned) as early as the 1950s. Archived internal notes show that they did not doubt the scientific evidence but decided to publicly deny it to keep their businesses going. The United States demonstrates impressively just how successful they were in planting seeds of doubt.

Fast forward into the twenty-first century. Global greenhouse gas emissions are still on the rise (the current temporary dip due to a world in lockdown does not change this picture). Climate change has evolved from a vague future threat to an everyday experience, albeit one that may not yet be recognized as such by everyone. Global mean temperatures of a degree above preindustrial temperatures and carbon dioxide levels in the atmosphere above 400 parts per million manifest as rising sea levels and changes in the frequency and intensity of some extreme weather events. These changes are not just making European summers uncomfortably hot. They threaten decades of development gains, and they pose a clear and present danger to the social and economic welfare of communities and countries around the world. While the global elite was busy ignoring or actively denying human-caused climate change, the problem worsened and devastating weather events proved the science to be correct. The price is being paid by those who always pay—people in developing countries, people who have to work outdoors, people who can't afford insurance—in short, people who have profited the least from improved living standards in a fossil-fueled society. And of course the price will be paid most by those who were not alive in the 1960s, '70s, '80s, and '90s, when those with influence chose to ignore climate change. It is the people who have no responsibility for causing climate change who are now taking to the streets, the courtrooms, and hopefully soon all the circles where decisions are made.

These are the facts and have been for many years, apart from the last point. But of course this point is crucial. Thanks to the young people on the streets, we are talking about climate change. And talking is the first step, as a problem that is not addressed cannot be solved. So these kids have achieved what scientists and activists could not achieve in decades; today we talk about climate change almost everywhere. And we do not only talk. Countries like the U.K. have adopted targets to achieve net-zero carbon emissions by the middle of the century; other countries and cities have declared a "climate emergency." This is a hugely important first step, but at the moment when it comes to climate, "emergency" is too often just a word. In order for climate emergency to mean anything it needs to be clearly defined and backed up with legislation. Instead we focus the discussions on individual actions (such as flying). But we cannot reach a net-zero target in thirty or forty years' time by holding individuals alone responsible. We live in a system built on the burning of fossil fuels and we need to change the system as a whole and do it quickly. Today's teenagers have already taken a lot of responsibility by telling the world what is at stake and that we need to act now, but we, the older people in powerful positions, need to implement the pathway to a carbon-neutral system. How to do that raises questions of responsibility—social, political, and philosophical questions. Changing our global society is not a scientific problem.

So why a book about a different way of doing climate science? Given we've known for a very long time that continuing to burn fossil fuels and emit greenhouse gases into the atmosphere is not compatible with stabilizing global mean temperatures, one could argue that the problem now passes to the rest of society. And to a degree this is certainly true, but even if we could achieve global net-zero emissions today, the world is already a degree warmer than it used to be. This has consequences, but we are only beginning to understand these consequences. The world is on the verge of largely accepting that climate change is a fact, but now we need to go from acceptance to understanding. We have known that temperatures increase over continents, but we do not live in continental averages, we live in cities and villages, tropical regions and arid zones, mountains and valleys.

It is in our local areas where we need to deal with the manifestations of climate change and where decisions are made on developing a region and implementing adaptation measures. If we are to reduce risk and build resilient systems, we must have the scientific evidence explaining our particular manifestation *when* this evidence is needed most—in the immediate aftermath of events, when key decisions are made about rebuilding, relocation, and recovery at a local level. And we must have it *where* it is needed most—in the most vulnerable regions, where events cause the largest impacts, where climate change is increasing risk, and where the media and the general public are asking questions about the causes of disasters and about their own vulnerability. The science described here, known as event attribution, allows us to link extreme weather events directly to human-caused climate change—essential evidence that has long been missing. The lack of such evidence has played its part in making it easier for governments and industry to shirk responsibility. Of course it would be better if we could have all this information before an event happens, and from a science point of view we can, but we are human, so we often only realize our vulnerability when it is threatened.

Attribution science can also be used to speed the societal transition up. For one thing, very straightforwardly, if every time an extreme weather event happens you hear an assessment of how much climate change is to blame for it, you are constantly reminded about the reality of it. If it's your grandmother who is admitted into hospital because of a much more intense heat wave, it will affect you differently than reading about a threatened polar bear.

Furthermore, attribution studies can be used to pin responsibility on those companies and countries that have profited most from not acting on climate change earlier. We have scientifically laid the groundwork for linking shares of carbon emissions sourced from industrial carbon producers to specific climate impacts. The implications are that we have the concrete evidence to take those carbon producers to court. For the time being this is not straightforward, but the science is already very clear. And lawyers are working hard on this, in Europe, in the U.S., and around the world. It is no longer a question of *if* a major carbon producer will be successfully sued, it is a question of *when*.

The links between vulnerability, climate change, and inequality are very real. Leaders of social movements state that climate change is a crisis of inequality. By demonstrating the degree to which human-caused climate change is responsible for disasters, event attribution may create new momentum for social movements.

In bringing climate analysis from the world of models and international policy into everyone's backyard, the courtrooms, and the streets, attribution is a science of responsibility. It is a responsibility that cannot wait until the generation striking today has grown up.

PROLOGUE

The New Weather

W E ARE THE first generation to experience a different type of weather. Weather that palpably deviates from what was encountered by our grandparents, their grandparents, their grandparents, and so on.

During my lifetime, the temperature on Earth has increased by around 0.6 degrees Celsius (1.1 degrees Fahrenheit), fundamentally altering our climate and thus the weather.* This change didn't announce its arrival; instead, it has slowly infiltrated our lives like a bad habit or physical pain. So far, in most of the world, we have responded with little more than a vague sense of unease.

Our unease comes from heat levels we believed to be the preserve of far-flung regions, torrential downpours that flood our streets and basements, and storms that uproot

---

* Climate is simply long-term weather in an area, often determined using thirty-year averages of weather observations.

huge trees and bring railroads to a standstill. Something has shifted in the fabric of the weather.

Throughout Europe, anxiety intensified in summer 2018. People faced relentless heat and merciless drought, heard farmers complain of crop failures, and hoped in vain for cooler temperatures that simply refused to materialize. In Greece at the end of July, following a series of forest fires, the famous Marathon Avenue in eastern Athens was lined with burned-out cars, charred trees, and windowless ruins. Later searches revealed the corpses of people who died in each other's arms, unable to escape the flames. Others fled into the ocean, six of them drowning.

As Europeans suffered in the heat, it began to dawn on many that perhaps climate change was not some distant threat but was making an impact here and now. And this experience was not confined to Europe. In Japan, heavy rainfall and flooding left hundreds of people trapped on their roofs in early July 2018. One year earlier, in September 2017, the Caribbean island of Barbuda fell victim to extreme weather when it was completely destroyed by Hurricane Irma and the entire population had to be evacuated to the neighboring island.

In September 2017, climate scientist Michael Mann of Pennsylvania State University declared that "I don't think it's a coincidence that during that period we've seen the strongest hurricanes globally,"[1] making reference to Patricia (Pacific, 2015), Winston (southern hemisphere, 2016), and Irma (Atlantic, 2017).

Nevertheless, quite a few people remarked that extreme weather has always existed. It is well known that our perception and memory become skewed as we age. The hurricane over the North German lowlands and the flooding of the River Elbe made it onto television thirty years ago, but back then it was rare to see reports of flooding in Bangladesh or heat waves in Kenya. In today's networked world, we hear of disasters in even the most remote regions of the world. Has weather become more extreme, or are we deceived by faulty perception and reporting frequency?

In many instances the answer is no, we are not being deceived. Why? Because we, the human race, have altered the parameters for our weather. From hurricanes to light summer rain, every weather event takes place under different environmental conditions than those of 250 years ago. Climate change does not only affect people in "developing countries." It will not be a phenomenon for our sons and daughters and their sons and daughters to tackle at some point in the future. Climate change is showing its colors now, and it's using the weather to do it.

What is deceptive is that it's really not that easy to determine whether a severe storm is within the range of "normal" and sheer bad luck, or whether storms that previously occurred every hundred or thousand years are suddenly becoming much more common. Despite the frequent newspaper headlines, the climate change we have set in motion cannot be blamed for every single weather event. When we ask whether the weather has become more extreme, the answer is that in many cases it has, but not always and not in every situation.

Determining the role of fossil fuels in all this requires scientific research—the research carried out by our small

World Weather Attribution team. When we founded the project in 2014, it was tantamount to a revolution in climate science. We reconstruct extreme events by analyzing weather data and comparing the results with weather simulations from our computer models. In just a few days or weeks, we can do something that for many years seemed impossible: we can attribute specific weather events to climate change or, on the contrary, prove that climate change was not involved at all. Our new field of research is therefore known as event attribution science. Instead of talking simply of general climate processes over periods of thirty years (as climate researchers always used to do), we focus on the things affecting us in the here and now.

For a long time, scientists actually frowned upon those who talked about the current weather. Our project is changing this perspective; for the first time in history, we have the means to make reliable statements about individual weather events. In a way, we are turning climate science on its head—even though we know it rubs some of our colleagues the wrong way. We want to replace unease and hazy gut feelings about the causes of weather with actual facts, something that hasn't been done before—and certainly not with such speed.

Media outlets have always reported immediately and extensively on storms, floods, and heat waves; it boosts their ratings. However, they almost always focus on the event and its consequences, nothing more. They rarely indicate that the weather event was unusual for the region or time of year. Most newspapers do not mention the specific area in which rain fell to trigger flooding, or whether this was an extreme—and therefore rare—meteorological event. Perhaps the rain itself wasn't unusually dramatic, only its impact?

Weather was (and still is) tolerated as though it were decreed by the gods, but we have long known that this is not the case. Today's weather is different because the human race has changed the climate, yet this fact is often drowned out by the melee of interests and ideologies. In principle, anyone can claim whatever they want; climate change deniers, energy industry representatives, and their political allies dismiss storms as the whims of nature. After all, we have always had terrible weather. As long as uncertainty reigns, it will be difficult to show a clear link between the weather and the mining and burning of coal, oil, and gas. Other people—including many evangelicals in the U.S.—see hurricanes as an act of God, as a punishment to be endured for the world's transgressions. Still others place all the blame on climate change. These are often environmental activists and scientists with initially good intentions to wake people up, to take drastic action to open their eyes to the dangers of climate change. As we know, however, there's a fine line between wake-up calls and alarmism. People so dedicated to the environment that they are willing to commit criminal acts are actually supporting politicians who hide behind climate change whenever their lack of or misguided planning turns a weather event into a disaster—it's all down to climate change, there's nothing we can do.

None of these perspectives are based on facts, facts that our new branch of science seeks to uncover. Over the last five years, we have repeatedly used our new method to discover whether and to what extent climate change is manifesting itself in our weather—in heat waves, droughts, and floods—with the aim of making climate science a thing of the present, not the future.

If everything goes smoothly, we can calculate the role of climate change in a weather event within one week—while it is still the focus of the media. It is crucial that we act in real time; this is the only way to influence the debate and to show people that climate change is not just a phenomenon of the future, that it is already happening before our eyes and in some of our backyards.

This new method also allows us to ensure that the world is better prepared for a changing climate. Only when we know which weather events are much more likely to occur in which seasons and regions can disaster management funds and efforts be deployed effectively. This knowledge could save lives.

It is also highly likely that our work will help call to account those responsible for the change in the weather. The future will see more and more energy companies summoned to court, and the first lawsuits based on natural disasters and climate change are already underway. With the aid of our attribution studies, corporations could be forced to pay compensation for climate-related damage to those without a lobby behind them.

Yes, facts can be powerful. They create clarity. I would like to use a specific event to explain how we, as scientists, arrive at these facts. I have chosen Hurricane Harvey, which swept across the southeastern U.S. in 2017, engulfing Houston in huge quantities of water. I find it to be a particularly good example not only to explain the basics of our work, but also to illustrate the insidious nature of lobbying, so that a climate policy is shaped by certain interests.

Day 0

It all begins with a stroke of fate. Unusually high ocean temperatures of over 30°C (86°F) have turned a tropical low-pressure area in the Gulf of Mexico into a cyclone. Mountains of cloud over half a mile high can be seen rotating counterclockwise around the center of the storm, driven by the Earth's rotation. The satellite images foretell not only extreme wind speeds, but also enormous amounts of rain.

The storm is constantly sucking new warm and humid air from above the warm water below, giving the cyclone more power. Meteorologists have upgraded the tropical storm to a hurricane, and it's rapidly making for the coast of Texas— Houston, to be exact. Houston is the fourth-largest city in the U.S., with almost seven million residents in the greater metropolitan area. With its many refineries, it is a major trading center for crude oil. Memories resurface of Katrina, the 2005 hurricane that killed more people in the U.S. than any in the last hundred years.

August 24, 2017. Our team is about to face this new event. It has already been named Harvey. Safe on the other side of the Atlantic, 4,785 miles away in Oxford, I fish my cell phone out of my bag and skim the reports while eating my breakfast. A tweet from American meteorologist Eric Holthaus catches my eye:

Just-completed GFS model (12z) shows nothing less than a flooding catastrophe for Texas. 24–48" of rain in 3 or 4 days. Please pay attention (@EricHolthaus, August 23, 2017)

We need to act, that much is clear. We need to determine the causes while all eyes are still on Texas.

We will influence what many people in Houston, the U.S., and the rest of the world will think about the hurricane. We will influence who or what will take the blame once the storm has passed.

Harvey is the first hurricane to hit a U.S. coastal city for years and will most likely plunge it into chaos. In one of nature's ironies, this is happening during the reign of a president who denies the existence of climate change and has announced that his country—historically the largest producer of greenhouse gases—will withdraw from the Paris Agreement. This would make the U.S. the only country on the planet to officially refuse to continually reduce its greenhouse gas emissions.

If we do not speak out and our team does not intervene, we will be yielding the stage to people with a solely political agenda who will make wild speculations based on their particular worldview. We will also be leaving a large section of the population under the delusion that the weather and climate are not connected—or at least that the connection is too complex to calculate.

Climate scientists themselves have encouraged this belief; with every storm that occurs, they have commented that a single weather event cannot be attributed to climate change. For a long time, it was not possible for climate researchers to speak about specific weather events, and for many it remains a taboo.

And yet all modern weather events take place under altered environmental conditions. After all, we have spent centuries burning fossil fuels, heating up our atmosphere by approximately 1°C (1.8°F) to date and changing the atmospheric circulation—where and how high-pressure and low-pressure areas move and develop. These days, every storm has something to do with climate change. The question is, to what extent? Has climate change weakened the storm or made it more powerful? Both are possible. And this is where we come in.

The problem with Harvey is that our team has never made these calculations for a hurricane. Hurricanes are complex and, as a physicist, I have a healthy respect for them. It is only since the start of the satellite era in 1979 that we have truly been able to observe them. In contrast to droughts and heat waves, they cover only a very small area, making it more difficult to simulate them in climate models.

I'm not sure whether we can do it, particularly in just a few days or weeks. If we hurry and miscalculate, we could jeopardize all the effort we have invested in the reputation of our (only just established) branch of research. Publishing academic studies involves many necessary yet time-consuming procedures; there are many risks to circumventing these procedures and publishing results while the world is still talking about the event.

And yet this is precisely what we need to do if we actually want to make a difference, to get involved in debates, and to put climate science on the offensive. With our approach we know we are treading on very thin ice that threatens to break at any moment. After all, we are (at least for a transitional period) breaking with the principle of peer review, established for centuries and one of the tenets of academic quality assurance. In the peer-review process, independent assessors from the same specialist field evaluate a study before its results appear in an academic publication. It makes perfect sense, and we would love to be able to take our time sounding out our analyses.

We do stick to the usual procedure, but only when developing new methods. In all other circumstances, we simply do not have the time. If we wait months to release our results, the general public will no longer be interested in exactly what happened. Other extreme weather events will probably have occurred in the meantime, and the public, the media, and politicians will have turned their attention elsewhere. We will have lost our audience.

And so we divide the peer-review process into two stages, well aware of the contradiction. However, we do not publish all possible figures indiscriminately as soon as they are calculated. We only use methods that have already been published in specialist journals and only present unreviewed results to the public if they relate to a new event, but not a new *type of* event. This may not be the academic custom, but our science is still good. So who are we?

Our Team

We are not police officers, disaster relief workers, or paramedics, and we are definitely not politicians. We are climate researchers, but we are no ordinary climate researchers.

Our core team is made up of myself, Geert Jan van Oldenborgh (also a climate scientist), and our manager, oceanographer Heidi Cullen. We all know each other really well. Even when nothing is happening, we hold a video-conference at least every two weeks; when we have a case, we speak almost daily. We have been doing this since 2014; Heidi from Princeton, Geert Jan from De Bilt in the Netherlands, and I from Oxford. We are assisted by many people, including my colleagues in the Environmental Change Institute at the University of Oxford, but we three are the ones who put our necks on the line. We decide when to take on a case, when an event has been sufficiently clarified, and when we go public. Just like today.

Our conference begins at 1 PM. By now Harvey has moved closer to the mainland and changed shape, and not for the better. Air force fighter pilots have flown over the storm and managed to identify the eye.[2] In the space of half a day, the cyclone in the Gulf of Mexico has drawn strength from the warm water and is now reaching maximum wind speeds of up to 99.4 miles per hour—compared to just under 62 miles per hour in the morning.[3] Governor of Texas Greg Abbott has announced that a state of emergency will be declared as a precaution so that swifter action can be taken if required. According to the National Hurricane Center, Harvey is set to reach the coastline late on Friday, August 25, or early Saturday morning.

President Donald Trump has also responded. He has posted images of himself at the Federal Emergency Management Agency (FEMA) in Washington, D.C., gesticulating as he talks with the agency chief. In the accompanying text, he warns the people of Texas to prepare for an emergency. The aim of all this is to show that, when it comes to the crunch, they can rely on their president. That he recently slashed disaster relief funds and denies the impact of climate change is another matter.

An ocean away from the storm, I am sitting in my little office in a brick building in Oxford. My team and I are trying to figure out what we can do. And whether we should start to investigate.

Nobody is surprised to hear that Geert Jan has already acquired the observation data and weather forecasts. Since Harvey made its way over the Atlantic from Africa's west coast and was highlighted as a potential hurricane by forecasters, he has updated the data almost every hour in the weather models of the Royal Netherlands Meteorological Institute, where he has been working for twenty years. Geert Jan pushes our group. If it were up to him, we would publish all our studies within two days. Sleep? Who needs to sleep when you could be analyzing weather data?

For us scientists, the world is to a large degree made up of numbers, diagrams, and computers in our daily work. However, we also need a connection to the event itself and to the damage and victims the disaster leaves behind. Disaster relief and aid organizations are our eyes on the ground; their job is to get people to safety, both before and during a disaster. The data they provide for us are difficult to obtain. How many people are affected? Has anyone been injured or killed? Teams in the area often know before we do whether a weather event could trigger a disaster because they receive early warnings from local weather services and other service providers. With all these data at their disposal, they can assess whether residents are prepared for the event. What they learn about Harvey is cause for concern: Houston's infrastructure is not equipped to deal with the extreme rain expected. And there is still no way of knowing whether the locals will take the warnings seriously.

We don't always agree on who to address most urgently. The disaster relief workers think it should be the local politicians because it is they—not the politicians in Washington or the media—who ultimately decide whether to rebuild houses or tear them down, whether to simply return to normal life or adapt their plans to the altered environmental conditions.

Heidi and her colleagues focus mainly on the media, aiming to reach people around the world. As well as organizing our team, Heidi acts as a knowledge broker, summarizing our specialist articles so that everyone can understand them. She believes it is important for us to share the key points of our work once public awareness grows.

Heidi is worried about Harvey. The pressure is immense. This is an important storm, so we can't stay silent, but hurricanes are new territory for us and we don't have any tried-and-tested methods or models to fall back on. So as not to make ourselves vulnerable, we must not under any circumstances make statements that are not 100 percent reliable. The U.S. attracts far more attention than the rest of the world, so we must take all the time we need.

"As soon as we speak, the public's attention will be on us," I say. "The whole world's press is watching Harvey."

"We have the observation data; we can concentrate on the rainfall rather than the storm," Geert Jan suggests, reminding us that we conducted a similar study one year ago for a region very close to Houston—a study on extreme rainfall in Louisiana in August 2016. In the case of Houston, we simply have to shift our geographical focus a little to the west. "It'll take less than two days if we use the American model; more time won't change the result."

"Yes, it will... maybe," I answer back, pointing out that this time we're working with a different circulation system. "Time allows us to find errors, to formulate our ideas carefully, and to apply several different models. Plus, the storm hasn't even finished yet—people will still be interested in the role of climate change in two weeks' time!"

Heidi agrees, making it clear that we have to be even more careful than usual—even though we know that every day that passes will give lots of people a platform to spread crude theories about why Harvey happened.

Slightly disappointed, Geert Jan leaves us with: "This isn't the last time we'll speak about the schedule..."

His discontent doesn't last long—analyzing and updating the observation data for Harvey is a truly exciting task. At least from Europe, safely removed from the storm.

My task within the analysis is the climate models and, above all, to ensure that the right simulations are available at the right time. This proves more complicated than expected. While we do have a regional climate model for the Gulf of Mexico over in Oxford—a model that covers the entire planet's climate system but simulates Central America in four times the resolution—the simulations were produced by a colleague in Mexico, which is why they are now stored on a server in the state of Baja California. In the twenty-first century it should be a breeze to transfer the data to

14

Oxford, but all I can do for the moment is send my colleague an email.

Our team decides that we need a hurricane expert, very few of whom exist. One is Gabe Vecchi, who has spent many years researching in the Geophysical Fluid Dynamics Laboratory at Princeton, one of the most important research centers for tropical storms. We worked with him a year ago when large swaths of Louisiana were underwater.

We also need someone with access to the latest local observation data—Antonia Sebastian of Rice University in Houston. Our team is completed by Karin van der Wiel, who sits next to Geert Jan in his office in De Bilt but has been working with Gabe in Princeton for a long time.

Experts are important, but you have to know and trust them before you can put them under time pressure—particularly in an international team such as ours. We've learned the hard way just how crucial trust can be.

By the end of our videoconference, our tasks are clearly assigned: Geert Jan will contact Gabe, Karin, and Antonia. I will speak to my colleague in Mexico and the IT team in Oxford as quickly as possible. We need to transfer several terabytes of climate data to Oxford; simulating hurricanes takes up a lot of memory.

As we wrap up the conference, our decision is made: Harvey will be the subject of a World Weather Attribution study.

A NEW BRANCH OF RESEARCH

The Role of the Climate in Our Weather

CHAPTER 1

CAUSE AND EFFECT

How We Created Our Weather

PEOPLE LIVING TODAY are among the first to clearly feel the consequences of a process that began 250 years ago in a Glasgow laboratory. James Watt, a Scottish mechanical engineer and inventor, concocted a new way of reducing steam and fuel consumption in fire-powered engines. When he and his steam engine paved the way for mechanical power and locomotives, they awakened humanity's voracious hunger for coal, oil, and natural gas. Since then, billions and billions of tons have been extracted and pumped from the ground to be burned in power plants and vehicles, all the while heating up the Earth like a greenhouse.

Why We Need Greenhouse Gases

The Earth draws its energy from sunlight. However, only some of the sun's rays reach the Earth's surface. Some of them—the UV (ultraviolet) rays—are absorbed by the ozone layer. Around 30 percent are reflected straight back into space from within the Earth's atmosphere or by the ice and other light surfaces on Earth itself. The rest of the sunlight is absorbed by the Earth, which then heats up and emits its own radiation. We cannot see this radiation, but we can feel it in the form of heat; this is mainly infrared radiation. Again, this is absorbed by the "greenhouse gases" in the air and reemitted in all directions with less energy. The most important greenhouse gases are water vapor, carbon dioxide (CO_2), and methane. Some of the radiation is thrown back to Earth, like a game of table tennis between the Earth's surface and greenhouse gases in which the energy decreases with every point of contact. The molecules absorb some of the radiation energy and convert it into kinetic energy. And that can mean only one thing: the atmosphere grows warmer.

Greenhouse gases make the Earth's atmosphere over 30°C (54°F) warmer than it would be without water vapor, carbon dioxide, and methane. Without greenhouse gases, most of the radiation emitted by the Earth would return to space unhindered, and our planet would be a pretty cold and uncomfortable place to live. We need greenhouse gases.

As long as the proportion of greenhouse gases and solar radiation remains constant, everything will be fine. The problem starts when we burn huge quantities of fossil fuels, releasing more and more greenhouse gases into the air that are able to absorb more radiation—greenhouse gases like carbon dioxide. Unlike water vapor, which leaves the atmosphere a few days after entering it in the form of rain, carbon dioxide sticks around for hundreds of years. To maintain the energy balance, the Earth has to warm up. We have known that this is happening since 1896, when Svante Arrhenius discovered the connection between global warming and greenhouse gases.

Since 1776, the year in which King George III granted James Watt patent number 913 for his steam engine, the Earth's temperature has increased by around 1°C (1.8°F). Carbon dioxide emissions increased slowly at first and were then accelerated by industrialization. Accordingly, the average global temperature rose gently at first by just 0.2°C (0.4°F) to 1960. Today, it is 1°C hotter worldwide. The hottest year on record was 2016, the second hottest was 2017, and the third hottest was 2015, followed by 2018, 2014, 2010, and 2013. The seven hottest years have all taken place within the last decade.

However, this additional 1°C on the global average is an abstract measure. We don't notice it directly, only its effects. To put it bluntly, the change in the average global temperature isn't killing anyone. At least not directly. But its influence on the weather is.

The Face of Climate Change

This 1°C has major consequences for our weather. Because the Earth's atmosphere is connected via global circulation, temperatures rise in almost every region on the planet. In the simplest case, the temperature rises everywhere, the probability of heat waves increases, while cold spells become less likely.

When the air becomes warmer, it can absorb more water vapor before the water begins to condense and form clouds.

The water remains in the air and clouds for a few days. But if the relative humidity exceeds 100 percent, it will fall as rain or snow. It's a simple formula: the more water the air absorbs, the more it rains. It's like a sponge; the larger it is, the more water it can soak up—and when you squeeze it, it releases all this water. Our atmosphere is like a constantly growing sponge.

The tropics demonstrate this well, as it generally rains much more intensely there than at temperate latitudes. But we can see the difference in Great Britain, for example, too; we simply have to compare the seasons—summer rain is often much heavier (though shorter and less frequent) than winter rain.

However, the Earth's growing hotter does not mean that we will all experience nothing but tropical downpours no matter where we are. The quantity and intensity of rain is only increasing as a global average; some places will get more rain, others less.

Rising temperatures and increased water vapor in the atmosphere follow simple physical laws. We climate scientists refer to the changes in weather that follow these laws as the "thermodynamic effect."

Climate change also influences the weather in another way.* Greenhouse gases don't just make the atmosphere warmer; altering its composition by adding more carbon

---

* Global warming is the increase in the atmospheric temperature and its consequences, while climate change also includes the effects of greenhouse gases directly. In other words, global warming is part of climate change, but climate change may manifest in many ways other than warming.

dioxide, methane, and water vapor (as we have done) also changes the atmospheric circulation.

Essentially, circulation is the movement of air that we experience as wind. It is created by the constant balancing of differences in pressure and temperature. If you've ever blown up a balloon and let it go without tying a knot in it, then you'll know that higher and lower pressure compensate for one another unless prevented by something like the shell of a balloon. Differences in temperature result from the fact that the Earth is more or less round and, therefore, the equator gets more sun than the poles because the sunlight hits the equator vertically and the poles at a sharp angle. This temperature imbalance creates wind systems—"jet streams"—that cover an entire hemisphere. They blow where cold and warm air masses meet and are diverted and accelerated by the rotation of the Earth. They blow all the time at high speeds and high altitudes.

And yet pressure and temperature also differ on a smaller scale: air heats up faster over land than it does over water, and faster over flat terrain than over mountains. Then we have clouds, which also influence temperature and pressure. If we change all these factors—the temperature, composition of the atmosphere, and cloud formation (and we are changing them)—then we also alter the circulation. We are therefore changing when and where areas of high and low pressure develop and where they go, when and where it rains, the strength of the wind, the season in which the wind blows, and the direction from which it comes.

Other factors also play a role, such as how the land surface is used and how it interacts with the atmosphere.

These changes have consequences. Today, hurricanes can develop in regions where this was not previously possible. The oceans are heating up, and in some regions the temperature threshold at which the water becomes so warm that it provides enough energy for a tropical cyclone to form has now been exceeded for the first time. For centuries, we have experienced weather in a stable climate—but with global warming, some of the familiar patterns of rainfall, drought, and storms are being disrupted.

Among climate researchers, this effect of changing circulation is known as the "dynamic effect." While it also follows physical laws, it is much more complicated to reliably simulate in climate models than the thermodynamic effect.

These two effects always occur together, never in isolation; however, since the dynamic effect in particular may vary wildly in strength and impact, the effect on the weather can differ greatly. If both warming and circulation change in the same direction, the overall effect will be stronger and disaster may be looming. Some regions experience more rain simply because the temperature of the air has increased and it can absorb more water. But when more low-pressure areas move into the region, bringing rain with them, the two can combine to produce an awful lot of rain.

My fellow U.K. residents know what I'm talking about. When I moved from Potsdam to Oxford a few years ago, I braced myself for typically rainy British weather, but it took a while to materialize. Over the last few winters, however, the south of England has experienced precisely this dual effect: More low-pressure areas from the Atlantic have brought more rain than would have been expected in pre-industrial times. In addition, the warmer atmosphere has made the rain heavier. The south of England has always experienced most of its rain in winter, while snow is rare. However, climate change has increased the probability of record rainfall such as that of January 2014, the wettest January since records began.[1] And it often doesn't stop with the rain: the more that rain falls, the higher the risk of flooding, particularly where houses have been built on floodplains. And there are plenty of these areas in southern Britain. Thanks to an ingenious flood system, Oxford was largely spared from floods that winter—unlike the people living farther south. Large swaths of Devon and Somerset in particular were transformed into lakelands. Railroads broke away and were undermined by the water, isolating these regions from the rest of the country for several weeks.

Equally, these two effects can take the wind out of each other's sails and have opposing impacts. While it may rain more on average in the warmer atmosphere, the altered circulation may mean that fewer low-pressure systems develop or that they enter a particular region less frequently. Ultimately, everything remains the same and the probability of wet winters or summers does not change, despite the altered climate. For example, the flooding of the River Elbe and the Danube in 2013 was not an extreme event that bore the hallmarks of climate change.[2]

There is, however, a third possibility: the change in atmospheric circulation is so strongly opposed to the thermodynamic effect that it wins out over the thermodynamic effect. In other words, the dynamics of when and where it rains change so much that certain regions suddenly receive hardly any rain at all. It doesn't matter that the warmer air could hold much more water vapor and produce a lot more rain; without the right flow of air, no rain clouds will form—and if the land is already dry, no clouds will form locally. This explains why the risk of drought can increase in some parts of the world even though the world as a whole is growing wetter. For example, southwest Australia has seen a dramatic decline in rainfall in the last fifty years; this is partially due to climate change.[3]

Looking at the world as a whole, it is fairly easy to explain how the climate can influence our weather. But of course, if a hurricane develops and threatens thousands of coastal residents, nobody will be interested in large-scale averages.

To date, nobody has systematically taken stock of the impact of human-caused climate change. To do so, we need to step away from large-scale averages and understand how climate change is reflected in a specific drought, flood, or storm. In other words, we need to make the connection between cause and effect, between where people live and what they are experiencing right now. This is now possible, but it requires laborious detective work.

Like all good detectives, we start not with the cause of an event, but with its impact. By finding out what happened.

Reconstructing the Event

It might sound trivial, but—as we know from every good thriller—it's not that easy to reconstruct an event. When it comes to floods, it is often not immediately clear where we should look at rainfall data. When floodwater breaches riverbanks, we first need to determine where the rain has actually fallen: Did it fall where the flooding occurred, or farther upstream? Was the rain actually heavy, or has someone simply failed to build a dam correctly? Or has the river been straightened, meaning that floodplains cannot prevent entire settlements from flooding? Was the entire season wet? Or was there an exceptional downpour on a particular day?

Most countries in the world have some sort of weather station network measuring temperature, rainfall, and air pressure every day. Measurements have also been regularly taken from satellites since 1979. Both of these observation methods document weather around the world and provide us with the data we need for our work.

Only once we have established the "what" can we begin to identify the causes. As a climate researcher, I frequently find myself grappling with seemingly mundane weather and not with the global climate. After all, a detective who investigates only the social causes of crimes will not be able to identify individual offenders—although the broader perspective may at least prove helpful in creating offender profiles.

Our cases are quite a bit more complex than those of the TV detectives; they are not scientists, and murders are usually committed by just one person. In contrast, every weather event has various causes and results from the interplay of local, regional, and global factors that differ every time. These may include very dry local ground, a volcanic eruption that blocks out the sun with a cloud of ash (temporarily altering the climate of a region or even the entire world), and climate change that impacts the whole planet.

None of these factors will trigger an extreme weather event by itself, and no weather event will occur again in the exact same form. To extend the metaphor, we are usually faced with a whole host of offenders, each extremely willful and with the unpleasant ability to change course whenever they fancy.

Yet some factors play a greater role than others. Climate change increases the probability of heat waves in the Mediterranean region. The "Lucifer" heat wave of summer 2017, which enveloped southern Europe in dry heat of over 40°C (104°F), provided our team with a particularly good example. Climate change has made this kind of sweltering heat at least ten times as likely; our best estimate suggests almost one hundred times as likely. If a heat wave like this were to be expected every hundred years without climate change, we can now expect it to happen every ten years or less with climate change.

We Define Extreme Weather

It may sound surprising, but there is no general definition of extreme weather. Whether an event counts as extreme depends greatly on the damage it inflicts on a region and how well prepared or vulnerable the region is—and where these vulnerabilities lie. In other words, there are no right or wrong definitions. There are only definitions relevant to the decisions that various regional planners must make to prepare for the weather of the future.

Like a sculpture, an extreme weather event can be viewed from multiple perspectives. Take the Serbian heat wave of summer 2012.[4] By defining the heat wave using summer temperatures, we calculated that it was made approximately ten times as likely by climate change. If, however, we define it as heat stress—incorporating the relative humidity—then it is only twice as likely. The humidity has therefore changed much less, and the heat stress (which takes humidity and temperature into account) is increasing more slowly than the temperature alone.

Absolute temperatures are likely to be crucial for farmers, who might be wondering whether they can continue to plant corn if June temperatures are more and more likely to exceed 40°C (104°F). Meanwhile, a cardiologist will focus on how heat affects the human body.

How we define an event determines the conclusions we reach. But just because a conclusion is different does not mean it is incorrect: for every definition, there is a correct answer. Given the usual inaccuracies of measurements and the fact that no climate model is perfect, we will only be able to find this answer with a certain degree of confidence, one that will be greater when we have good models and high-quality observations. So while there is no single true and correct definition of extreme weather, there are meaningful and less meaningful definitions depending on how important various aspects of a weather event are for our lives. If we want to determine the role of climate change, we must therefore begin by clarifying what extreme weather actually is and why the event is important for us. In the case of Hurricane Harvey, we began by asking ourselves these same questions: What kind of storm is it? Which of its characteristics are important for us? Let's get to work.

Day 4

On Monday, August 28, 2017, a sea breeze blows through the nave of a small church in Rockport, Texas, a town of ten thousand residents on the Gulf of Mexico named for the rocky overhangs at the waterfront. The community was supposed to be holding a weekend fair here. But half the church is missing. After making landfall on Friday evening, Hurricane Harvey swept over this idyllic coastal town with a mild climate and ripped the building apart.

Inside the church, wooden and steel struts stick out in all directions. The tiles are missing from the roof and foam spills out of the concrete blocks seemingly dumped in front of the church. Gusts of wind up to 130 miles per hour have shattered the houses of Rockport and brought rain that has turned freeways into lakes. Boats have been washed ashore, power lines have snapped, and the power supply is down. Dozens of residents have been injured and one has lost their life.

In the meantime, the U.S. National Hurricane Center has downgraded Harvey to a tropical storm. But Harvey simply alters its strategy of destruction, coming to a virtual standstill and lingering over the coastal city of Houston. In the space of three days, the storm releases more rain than has ever been recorded here in such a short period.

There is no end in sight, at least not a swift end. Moist, warm air over the ocean continues to provide the storm with energy, with no inland wind to drive it away. Already very extreme, this weather event is well on its way to escalating even further and becoming a "black swan"—a term used by meteorologists for a highly unlikely, unprecedented, but not impossible event. An event expected to occur every ten thousand years or even more rarely.

It is Day 4, and we still do not have a complete event—it has begun, yet it persists. Harvey is not yet history and continues to draw the world's attention. In fact, the attention grows day by day as the storm becomes a hurricane, then a meteorological phenomenon, then a disaster for Houston. The U.S. media are already beginning to speculate on the causes. Most of the headlines still have question marks: Is it climate change that has made Harvey so terrible?

However, some media outlets have already committed themselves—some speak of a clear connection,[5] while others (like Fox News) broadcast entire programs without mentioning the words "climate change" even once.[6]

This puts us under greater pressure. The world is not waiting for us—except for Harvey, who definitely seems to be waiting over Houston. We need a plan and we need it fast, even though the event has not yet finished. At least we have a relatively reliable forecast.

First we need to decide which of Harvey's characteristics we can consider. It would of course be best to analyze Harvey as a whole and to make its wind speed the basis of our study. But we have never analyzed hurricanes before, and with good reason: hurricanes are complicated, and to this day it remains virtually impossible to realistically and reliably simulate all the key variables in climate models. The observation data have only been viable since the satellite era. And the climate models that exist are not necessarily suited to simulating this kind of extreme weather.

We therefore need to find a compromise between what is actually happening and what we can reasonably research.

Rather than the actual storm, we concentrate on the rain it is causing and use this as the basis for our analysis. In other words, we ignore the storm and consider only the rain it brings. After the experiences of these first few days, this approach seems entirely appropriate: while the storm itself has undoubtedly caused damage, it is becoming increasingly apparent that it's the immense quantities of rain submerging Houston that are turning Harvey into a disaster.

The scenario is ideal for scientists like us. For this type of extreme rainfall, we have the right models and sufficient weather data dating back over one hundred years. One year ago, we studied rainfall in a region of Louisiana just 199 miles from Houston.[7] Here too we analyzed solely rainfall, regardless of whether it was triggered by hurricanes, tropical low-pressure areas, or other meteorological processes. After all, when the focus is on the damage meted out by such extreme rainfall, the meteorological causes are irrelevant.

Restricting ourselves to rainfall does not render our work trivial. This pragmatic step speeds up our search for the cause—or helps it seem a little less impossible.

So our work can begin. But the world wants answers now. In the early afternoon, I receive a call from Karl Mathiesen, editor-in-chief of *Climate Home News,* one of the world's leading climate news portals.

He asks me to write a guest article so that a scientific voice can be heard amid the sea of information and sensation surrounding Harvey. I now find myself in the exact same position as my climate research colleagues who have to give statements about extreme weather while lacking sufficient facts to draw conclusions for a specific event—and whose unsatisfactory answers I have definitely criticized in the past. I demur, saying that I am not a hurricane expert and have no concrete results. But Mathiesen doesn't let up, and in the end I agree. I want to counter both the alarmist tone of some of my colleagues and the message that weather events cannot be linked to climate change.

I write down what we are currently able to say about Harvey.[8] When the oceans become warmer, the conditions improve in which hurricanes can form. Warmer water provides the cyclone with more energy and gives it more power. However, climate change also influences hurricanes in another way: in a warmer atmosphere, the differences in vertical wind shear increase. This means that the horizontal wind on the underside of the storm moves at a totally different speed from the horizontal wind further up. If this difference is greater than 32 feet per second, the cyclone that forms the hurricane cannot keep going and the hurricane loses power. Climate change strengthens both the ocean temperature and the vertical wind shear—the main reasons why hurricanes develop and collapse.

In a specific case, it is therefore not immediately clear whether a warmer climate makes a hurricane more likely. Or whether Harvey and its deluge of rain are a taste of what Houston can expect in the future—only a targeted research study can answer that one. I also point out that we expect more intensive rainfall in a warmer climate. How much more depends on how the circulation changes. And that's what we're working on. We want to provide clarity, to help counter the powerful interests that have spent years trying to do the exact opposite: cause confusion.

CHAPTER 2

SOWING THE SEEDS OF DOUBT

Climate Change Deniers

IT DIDN'T TAKE Barbara Underwood long to settle into her new role as attorney general of the state of New York. Six months in, on October 24, 2018, she took on possibly the biggest opponent of them all, filing a lawsuit against the world's largest market-listed oil company.[1] Based on three years of investigations, the ninety-seven-page indictment alleges that ExxonMobil spent decades deliberately deceiving investors and the public about the dangers of climate change. It states that while ExxonMobil has long known about the consequences of global warming and has even helped to drive research in this field, it continues to downplay the risks of climate change to the outside world—and that this is not classified as freedom of speech.

"Investors put their money and their trust in Exxon," said Underwood.[2] She asserted in a statement that ExxonMobil "built a facade to deceive investors into believing that the company was managing the risks of climate change regulation to its business when, in fact, it was intentionally and systematically underestimating or ignoring them, contrary to its public representations."[3]

There is a lot at stake for ExxonMobil; it could be sued for hundreds of millions of dollars and its image could be seriously damaged.[4]

Al Gore (Nobel laureate, former vice president of the United States, and part of the alliance of plaintiffs) compared the case to the 1990s action against the tobacco industry, which had denied the risks of smoking for decades.

It may be time for the oil company to pay for the strategy it has pursued for almost four decades.[5] At the end of the 1970s, a serious problem was revealed: Exxon (as the corporation was known until its 1999 merger with Mobil) had commissioned scientists to conduct studies focusing on the link between oil production and "climate change," a process that had recently become a hot topic.

On August 24, 1982—exactly thirty-five years before Harvey was upgraded from a tropical storm to a hurricane and we began our study (which would also link indirectly to ExxonMobil)—a meeting was arranged for 9 AM at the Texas headquarters of the multinational oil company. Andrew Callegari, an engineering mechanics specialist at New York University, had prepared a presentation for the management with two "Topics to Be Discussed":

- CO_2 greenhouse effect
- Corporate research climate modeling[6]

Callegari went on to explain how burning fossil fuels heats up the planet. In addition to higher temperatures, rainfall patterns could shift and coastal sea levels could rise. He also included a forecast for Exxon's representatives (including some from the PR department): the first effects of climate change would make themselves known by the year 2000. Assuming Callegari was referring to our weather, he was not far off.

The corporate representatives had heard similar things from other scientists, but never so explicitly. What this university researcher was telling them was tantamount to an existential threat. A threat to the livelihoods of many people, but also—and this may well have caused greater concern at the company headquarters—to its business model. The company had to take action.

Three years previously, Exxon had begun to trial a strategy that it would stick to for decades. It would bring scientists like Callegari into the company fold, allowing them to research the consequences of burning oil and gas for the global climate and develop hundreds of studies together with other scientists. This would establish the company's image as a credible partner in upcoming discussions with legislators and governments. But these studies, analyses, and internal papers were just the start. The corporation took a two-pronged approach and launched a publicity campaign with one goal above all else: to spread doubt.

In addition to commissioning studies that largely acknowledged human-caused climate change, the corporation placed advertisements in newspapers. The most important appeared in the *New York Times*, the leader of the American press. From 1979 to 2001, Exxon paid to advertise every Thursday at the discounted price of $31,000 per ad.

However, the messages published did not reflect what Callegari and the other in-house scientists were saying in their studies. A typical advertisement from 1997, shortly before the Kyoto Protocol was adopted and industrial countries committed to protecting the climate, stated: "Scientists cannot predict with certainty if temperatures will increase, by how much and where changes will occur... Let's not rush to a decision at Kyoto. Climate change is complex; the science is not conclusive; the economics could be devastating."[7]

Again and again, these advertisements speak of gaps in knowledge, of a high degree of uncertainty, and unproven theories. And yet scientists reached a consensus in the early 1990s at the latest that human-caused climate change was already in full swing, could not really be reversed, and required major international effort to halt its progress.

In 2017, two Harvard scientists worked their way through 187 ExxonMobil documents and examined the company's climate communications between 1977 and 2014. They write that "the predominant stance taken in ExxonMobil's advertorials is 'Doubt.'" Given the discrepancy between the communications and studies, they conclude that "ExxonMobil misled the public."[8]

For a long time, businesses focused almost exclusively on traditional lobbying, ignoring public opinion. However, corporate leaders—particularly in the tobacco and oil industries—realized that it can sometimes be much more effective to influence public opinion and shift the perspective on controversial topics. Access to power is of little use if politicians striving to stay in office value polls and media reports more highly than the opinions of economic insiders.

Alleged Experts and
Conservative Think Tanks

In the case of climate change, the main strategy employed by energy corporations like ExxonMobil is to cast doubt on the value of environmental regulations. They do this by bombarding the public with wild claims: Earth isn't getting warmer—and if it is, maybe that's a good thing. Or at least that it's not caused by humans, but the result of increased solar activity. If you repeat such falsities long enough, at some point they will take root in people's minds.

These claims are most effective if they are made not by the corporations themselves, but by alleged experts who are being paid for their services. A study by researchers at the University of Central Florida states that "the central tactic employed by CTTS [conservative think tanks] in the war of ideas is the production of an endless flow of printed material ranging from books to editorials designed for public consumption to policy briefs aimed at policy-makers and journalists, combined with frequent appearances by spokespersons on TV and radio."[9]

Since the early 1990s, think tanks like the Competitive Enterprise Institute and the Heartland Institute have been leading the movement against global climate protection. In a television advertising campaign, the Competitive Enterprise Institute claimed that glaciers are growing, not melting, and had the following to say about carbon dioxide: "They call it pollution. We call it life."[10] A few years ago, the Heartland Institute attempted to have global warming removed from public school curricula.[11] It also claimed that global warming is natural and compared climate scientists to mass murderers.[12]

Armed with millions of dollars from oil corporations and conservative foundations,[13] this legion of climate change deniers has succeeded in drawing a disproportionate amount of attention to a small number of paid, often nonspecialist "experts" who hold a minority view within climate science. In the U.S. at least, this is a huge movement, not just the preserve of a few crackpots who happen to be in government. This isn't just a dispute between a few climate scientists and a few diehards; this is a cultural struggle dividing the U.S.[14]

Climate change deniers have more power in the U.S. than anywhere else. They are deeply rooted in the conservative movement that developed in the late 1960s in opposition to the peace and civil rights movements and now focuses mainly on opposing abortion, gun control, and the welfare state. The fight against national environmental laws and global climate regulation blends in seamlessly, fueled by fears of state interference in personal lives, and of losing sovereignty, economic strength, and the domination of the global North in the distribution of wealth and power.

Considerable numbers of people in the U.S. and other countries reject the basic findings of climate science because they conflict with their worldviews or their core conservative beliefs—at least according to Stephan Lewandowsky.[15] A psychologist based at the University of Bristol, he has been studying this topic for some time and sees the denial of climate science findings as the brain's defense mechanism to protect one's own identity.

After the Intergovernmental Panel on Climate Change (IPCC) was formed in 1988 and environmental protection became a global phenomenon at the 1992 Rio Earth Summit, the conservative movement organized a large-scale counter-movement with climate research as its mortal enemy. As the University of Central Florida researchers state in their study, "the sceptical position is strongly aligned with conservatism and the economic interests it represents."[16]

By striving for balanced reports and equal scrutiny of both sides, journalists actually play into the hands of those seeking to manipulate opinion at the expense of facts. Treating the existence of human-caused climate change as a controversy rather than a fact provides a platform that climate change deniers should never be given.

This doubt has even infiltrated the established, left-leaning media—particularly since the 2008 financial crisis, when many journalists, including science editors, were made redundant. At CNN, for example, responsibility for climate topics suddenly fell to weather presenter Chad Myers. In December 2008, he declared: "You know, to think that we could affect weather all that much is pretty arrogant."[17]

The U.S. is not alone in this issue. In Great Britain too, renowned scientists have increasingly found themselves sharing a stage with deniers whose only qualification is to have an opinion. In March 2018, the BBC was officially reprimanded for repeatedly providing a platform to notorious climate change denier Nigel Lawson without contradicting his lies on climate change or making clear that he was in no way an expert. In doing so, the BBC broke the rule that "news, in whatever form, must be reported with due accuracy and presented with due impartiality."[18]

The German media are not immune to doubting climate change either. After Fritz Vahrenholt, former Hamburg environment senator for the Social Democratic Party (SPD), published his book *Die kalte Sonne* (The cold sun)—which challenges human-caused climate change—the *Bild* newspaper published an article entitled *Die CO2-Lüge* (The CO_2 lie). And this was not an isolated case.

After many years, environmentalists and climate researchers have also learned some lessons and have succeeded in prizing open the united front presented by energy corporations as they reject environmental protection. With negative campaigns and a call for boycotts, they have convinced some managers that there can be advantages to considering the environment. How? By taking aim at a company's most important currency: trust. If its public image is damaged, then its business will be too. And that can bar it from access to politics.

Following the ratification of the Kyoto Protocol in 1997, multinational oil companies like BP and Shell withdrew from the network of climate change deniers, started to invest millions in green energy, and publicly announced that they would be reducing their greenhouse gas emissions; meanwhile, Exxon continued to brace itself against any form of change. Leaked government documents revealed that the corporation even persuaded President George W. Bush to turn his back on the Kyoto Protocol.[19] Writing in the *New York Times* a few years ago, Paul Krugman (Nobel laureate in economic sciences) described ExxonMobil—not wholly inaccurately—as an "enemy of the planet."[20]

Since then, however, even ExxonMobil has changed course—with some delay and with pressure from the public and its investors. After Rex Tillerson became chief executive of ExxonMobil in 2006, he canceled all direct contributions to climate-change-denying think tanks, invested in green energy projects, and even called for a carbon dioxide tax. This (largely cosmetic) realignment removed the company from the firing line of the international climate movement and allowed it, if not to obstruct climate protection measures, then at least to help shape them and even tailor them to its needs.

At least officially, energy corporations have dropped the line held by many conservative think tanks, which continue to seek confrontation. This explains why, in his brief spell as Trump's secretary of state, Rex Tillerson was one of the few cabinet members who did not want to withdraw from the Paris Agreement.

Let's make one thing clear: the world needs energy. And until we have sufficient renewable energies at our disposal, we will not be able to do without oil and gas, alongside nuclear power. But the equation is simple: to restrict global warming to 2°C (3.6°F), to keep climate change within a range to which we can perhaps still adapt, we must not exploit and burn more than one-third of known fossil energy reserves by the middle of the century (according to forecasts by the International Energy Agency).[21] With the Paris Agreement, every nation on Earth committed itself to reducing its greenhouse gas emissions to net zero in the second half of the century.[22]

American climate change deniers have attempted to expand their influence in countries like Germany too. At its annual conference on November 9, 2017, at Düsseldorf's

Hotel Nikko, EIKE (European Institute for Climate and Energy)—an association that rejects climate change—featured as its guest speaker Marc Morano, perhaps the most offensive manipulator of information on climate change. Now a PR manager, Morano was once a spokesperson for James Inhofe, then chair of the U.S. Senate Committee on Environment and Public Works. He now works for the Committee for a Constructive Tomorrow, a climate-change-denying think tank based in Washington with a subsidiary organization in Jena, Germany. To applause from a very gray and male audience, Morano proudly declared that the U.S. is the only country in the world to reject the "21st-century witchcraft" of the United Nations.[23]

EIKE is also supported by politicians from established parties, including the aforementioned Fritz Vahrenholt, who once worked for Shell and RWE (the German electric utilities company). Denying climate change is not the preserve of the far-right Alternative for Germany (AfD) party; such views can also be found in the Christian Democratic Union (CDU). The right wing of the CDU, which has dubbed itself the "Berliner Kreis" (Berlin Circle), declared in June 2017 that the causes of climate change are not yet understood, but that it was unlikely that the greenhouse effect played a solitary role.[24] Led by CDU parliamentary representatives Sylvia Pantel and Philipp Lengsfeld, the group claims that the climate has always changed and that the greenhouse effect is just one of several contributing factors along with solar activity, the Earth's position relative to the sun, volcanic eruptions, and meteorite impacts. According to them, the consequences of climate change are "far from proven." This reflects the earlier rhetoric of German energy companies: as recently as 2006, in a legal battle with Greenpeace, RWE described climate change as the "subjective perception of an assumed threat that is neither concrete nor current."[25]

Today, energy company executives have tailored their language—without giving up their business model, of course. They are simply shrewder, speaking less about climate change and more about jobs and business locations. And it's proving a great success: in contrast to countries like Great Britain, Germany's coal-fired power plants have continued to operate without restriction for many years, despite the country's energy transition. Today, the coal industry employs just tens of thousands of people—and while each job is undoubtedly important, it hardly justifies allowing climate change to advance unchecked. There's also the matter of continuing to employ people in an industry whose time is running out. Is that really the right thing to do? After all, nobody is seriously asserting that the coal industry has a future. After years of waiting and hoping for the best, in early 2019 a German coal commission set out a plan to decommission the last coal-fired power plant by 2038 at the latest.

Energy companies and conservative climate change deniers may have enjoyed long-lasting success—particularly in the U.S.—in preventing the public from seriously reflecting on the real dangers of climate change, but their strategy has now reached its limits. Thanks to the weather.

As long as climate change was seen as an abstract phenomenon, it was easy to portray climate science in a bad light and obstruct climate protection laws using false information. Now, however, there is no denying that something is changing on our planet. We can see and feel the consequences of global warming. Heat waves, floods, and droughts are robbing people of their homes, their jobs, and even their lives—and not just in developing countries in Africa and Asia, but also in industrialized nations like the U.S. Whatever the country, the poorest suffer the most; they live where housing is cheap, but also precarious, and feel the effects of even slight climatic changes.

The impact of climate change on our weather can no longer be denied. However, the idea that this impact can be proven in individual weather events is new and nowhere near as indisputable as climate change itself. This explains why the weather is the perfect setting for a proxy war between wide-ranging interests—a war in which energy companies and conservative movements bring their political allies into play, just like the environmentalists on the opposite side. Every cyclone and heat wave has become a battlefield, both sides contriving scenarios of conflicting extremes that wildly underestimate or overestimate the role of climate change depending on their beliefs.

Climate Researchers in the Crossfire

Of all people, climate scientists have barely entered the fray. Fossil fuel energy companies and climate change deniers probably couldn't believe their luck that the one group of people who could have stopped them from putting their own spin on cyclones, heat waves, and floods were silencing themselves by not commenting on the weather.

There were reasons for this cautious approach. In the U.S., climate researchers have been pilloried, attacked, and discredited in the media. James Inhofe, an eighty-four-year-old Republican from Oklahoma who (with the exception of a few years) chaired the U.S. Senate Committee on Environment and Public Works from 2003 to 2017, repeatedly invited climate scientists to hearings only to have them engage in discussions with supposed experts who denied the existence of human-caused climate change. Inhofe even called science fiction author Michael Crichton as a key Senate witness against climate change.

The campaign against climate scientists peaked in November 2009, when persons unknown hacked the University of East Anglia's server and made more than one thousand private emails and another thousand documents from scientists available online. Quotes were taken out of context and used to stir up a supposed scandal that was dubbed "Climategate," giving the impression that researchers had made climate change seem more dramatic that it really was.

The scientists were later acquitted by several institutions, which confirmed that there had been no scientific misconduct. But the damage had been done—just before the Copenhagen climate conference, which was supposed to be a breakthrough for a global climate contract but failed spectacularly.

Many American climate researchers have thus grown used to censoring themselves and carefully weighing every word. My generation of scientists is the first to have experienced "Climategate" only through newspaper reports and the recollections of older colleagues. We have a more relaxed relationship with journalists and the public—with exceptions on both sides, of course.

Many climate scientists still try to avoid public statements, preferring to conduct discussions in the safe environment of scientific journals, a world totally incomprehensible to most laypeople. These scientists restrict themselves to repeating IPCC statements when called upon to speak publicly, and with good reason. The IPCC has succeeded in doing what climate change deniers have spent a lot of time and money trying to prevent: since it was founded in 1988, the IPCC has published regular overviews of the status of climate change research, including causes, effects, and possible solutions. It is important to note that all IPCC authors are scientists working outside their regular jobs unpaid, and that the reports include only statements verified by multiple studies in specialist journals. Contradictions between research results are mentioned. What makes the IPCC reports different from all other summaries and overview articles is that they must be reviewed by the government representatives of the 195 countries in the IPCC and approved in a meeting before publication. This ensures that the statements made in the reports are as scientifically watertight as possible to provide government representatives with a sound decision-making basis. The crucial factor is that the IPCC does not make policy, and its reports do not contain any specific political proposals. The IPCC is a unique body that represents science but is above science. So there's a lot to be said for sticking closely to what the reports say.

However, the whole process—from determining the contents to publishing the report—takes around seven years. These reports have not yet said anything about specific weather events. This makes scientists particularly cautious when it comes to discussing the weather; they would be leaving their safety zone and opening themselves up to attacks.

But times have changed. The most recent IPCC report from 2013 was the first to state that it is now possible to attribute specific weather events to climate change. In 2015, through the Paris Agreement, every country on the planet acknowledged that climate change has already caused damage and loss, damage largely caused by extreme weather, as the specialist literature shows.

This is why our team wants to put climate science on the offensive, rather than the defensive. We can state whether and to what extent climate change is manifesting in our weather. We can stand up to the energy companies and mercenaries of doubt.

We already have a complete inventory of greenhouse gases, of how much has been emitted when and by whom.[26] Scientists are constantly updating this inventory on the causes of climate change, allowing us to work out a country's or company's historical contribution to greenhouse gas emissions. We also know what these emissions mean for the average global temperature—and can express it in very precise figures. For a long time, making specific statements on the effects of global warming required a great deal of maneuvering.

Today, we can complete the chain of causal evidence and create a basis on which to hold oil giants accountable and fairly distribute the burden of climate change. If we are quick enough, we can participate in the discussion while an extreme weather event is still occurring and cut short the proxy war (both sides are as irrational as they are emotional) with facts that provide clarity and carry explosive political power.

Day 5

It's raining. Unbelievable quantities of water are engulfing the city of Houston. Entire streets and hundreds of thousands of houses are submerged in brackish water. No other storm in the history of the U.S. has ever brought so much rain.[27] The Houston NWSO (National Weather Service Office) weather station measures 41.07 inches in three days. For comparison, my hometown of Kiel in northern Germany (which doesn't exactly lack rain) gets around 27.5 inches per *year*; Washington, D.C., gets 40.8. To adequately illustrate the quantity of rain, the U.S. National Weather Service has to add two new shades of purple to its weather maps.[28]

It almost seems as though these clouds were magically drawn here, to release torrents of rain and turn the world's attention to the place where U.S. oil production began more than one hundred years ago,[29] where the soil conceals one of the world's largest oil reserves. At the start of the twentieth century, just 124 miles east of Houston, Anthony Francis Lucas stood on a hill. An engineer and Austrian migrant, he had spent weeks using a steam engine to drive mud into the earth and coax out the black gold he believed it to contain. His efforts seemed in vain until January 10, 1901, when an explosion shook the shaft and mud gushed from the well seconds later, followed by a black-green jet shooting out of the drill tower some 160 feet into the air.

Tapping the largest oil field in the U.S. ushered in a new era. Oil production soared throughout the country with the Gulf Coast as its center. Soon 285 drilling sites peppered the Texas hill known as Spindletop, and many new oil firms came to loggerheads. One of those was the Humble Oil Company, known today as ExxonMobil.

It would be truly symbolic if we were to establish that climate change had played a part in the torrent of rain over Houston, so close to the birthplace of perhaps the most powerful oil corporation in human history, the oil corporation that has done more than any other to sabotage climate protection measures. A freakish link between cause and effect.

The continuing rain in Houston prevents us from saying whether this is a hundred-year flood or thousand-year storm. To be absolutely certain and not rely on forecasts, we need to wait until the event is over. But that doesn't mean we have to sit and twiddle our thumbs. We look for computer models capable of simulating the rain in this region. One key climate model, a computer simulation of the atmosphere, is stored on the server in Mexico, and I'm still waiting for a response to my email. We are also waiting for the Geophysical Fluid Dynamics Laboratory at Princeton University to permit us to use another model. We've already tested one model successfully. If we can do the same with a second, then we'll be much closer to a study.

While we are waiting to determine whether climate change is partly to blame for the downpour in Houston,

ExxonMobil goes on the offensive: CEO Darren Woods announces that the company will be donating half a million dollars to the Red Cross to support the rescue work in Houston and the people living on the Gulf Coast. "Our thoughts and prayers are with the residents of Texas and Louisiana Gulf Coast communities currently in the path of Hurricane Harvey," the company declares in a press release. "We hope our contributions will help provide comfort to our friends and neighbors in areas impacted by the storm."[30]

Among others, the Heartland Institute—the think tank that received long-standing payments from ExxonMobil—issues a press release: "In the Bizarro world of the climate-change cultists—though it has been nearly 12 years since a major hurricane has hit the United States—Harvey will be creatively spun to 'prove' there are dire effects linked to man-created climate change, a theory that is not proven by the available science."

It continues: "Facts do not get in the way of climate-change alarmism, and we will continue to fight for the truth in the months and years to come. But this weekend, our focus and our prayers will be with the people of Texas."[31]

And yet the media barely mention climate change. A few days later, an analysis by the nonprofit organization Media Matters for America will show that ABC and NBC, two of the three major broadcasting companies in the U.S., did not mention climate change at all.[32] Only CBS discussed a possible link between Harvey and global warming and allowed climate scientists to have their say. These scientists pointed out that the oceans are warming up—and therefore providing more fuel for hurricanes like Harvey—but also that relative humidity is increasing, which can lead to greater rainfall.

Some climate scientists have already voiced their opinions. "There are certain climate change–related factors that we can, with great confidence, say worsened the flooding," declares Michael Mann of Pennsylvania State University. "Harvey was almost certainly more intense than it would have been in the absence of human-caused warming, which means stronger winds, more wind damage and a larger storm surge."[33]

There is already speculation that Harvey may have caused $190 billion in damage, making it the costliest storm in the history of the U.S.[34]

But is it only climate change that has made it a disaster? After an initial analysis of the observation data from the weather stations, we are very certain that this is a truly extreme event. In the last hundred years, this region has never seen anywhere near as much rain in such a short period. With the aid of statistical models, we are now at least able to say that a weather event like this is so unlikely that it is to be expected less than every nine thousand years.

CHAPTER 3

REVOLUTION IN CLIMATE SCIENCE

Turning the Field on Its Head

I ALMOST MISSED ONE of the most important meetings of my career—a meeting that would launch our project to attribute extreme weather to climate change in real time—and for the stupidest of reasons.

On a sunny day in December 2014, I sat in a Starbucks in San Francisco with Myles Allen, my boss at the time, looking out onto the left side of Fourth Street. We each held a cup of coffee that was slowly going cold. Heidi Cullen should have been here half an hour ago. Our curiosity had been piqued when she contacted us to say she had a proposal. And so we had absented ourselves for a few hours from the world's largest conference of geosciences and climate sciences, which every December saw climate researchers descend on San Francisco.

Although she has a doctorate in oceanography, Heidi is not part of this group and is not a climate scientist. She was working for the nonprofit organization Climate Central in Princeton, which aims to provide television weather presenters with information on climate change that their audiences will understand. Back then, climate researchers were still giving the topic of "weather" a wide berth, so it stood to reason that she would ask my boss for a meeting and make a special journey to San Francisco.

In 2003, Myles had published an article in the high-impact scientific journal *Nature* in which he described for the first time how extreme weather events could be attributed to climate change.[1] He followed through with the idea one year later when he and colleagues from the British Met Office and Oxford studied the European heat wave of 2003, concluding that climate change had made it twice as likely. Myles had invented the method for attributing weather events—but it was Heidi who would help us to turn traditional science completely on its head.

Wondering whether to leave or order another coffee, we finally noticed another Starbucks a little farther north on the other side of the street. Rushing over, we saw Heidi's blond hair through the window. The overwhelming number of Starbucks branches in American cities had almost cost us the first meeting of our branch of research!

American through and through, Heidi turned on the charm and lavished us with praise for our pioneering work. Then she asked the crucial question: Could we see ourselves working a little faster?

That's not quite how she put it, of course. But that, essentially, was what she wanted to know.

Heidi and her team had been advising weather forecasters and journalists for years, and their way of working had
long since become second nature. The weather is constantly
being predicted. Every country has its own weather service. Private weather services sell their forecasts to radio
and TV stations but also to insurance companies, electricity
providers, and hedge funds. The latest weather report is
published as soon as it is ready. Nobody demands that each
forecast be scrutinized at length to determine its scientific
quality. Weather forecasts are not a scientific activity, but
climate research is. Although weather services and climate scientists use the same models, weather services do
the exact same thing every day, while climate scientists ask
many distinct questions, constantly use new and different
methods, and find that the answers can differ greatly when
it comes down to the details. In short, we conduct basic
research and weather companies perform a service, always
based on the same principle. Now Heidi was relying on us
to speed up the normally lengthy route from research to
operations.

As we drank our second coffee in San Francisco, she
at least seemed confident that we could bend our work to
meet journalists' schedules without losing the support of
the scientific community; the latter would indeed prove
challenging.

But on that December day, we let ourselves be won over
by her idea, by the concept of "World Weather Attribution,"
even though it would be radically different from the way
that scientists usually worked. However, it would be some
time before the project bore this name and really took off.

Scrutinizing a Core Principle

The peer-review process is one of the most important elements of science. Before articles are published in scientific journals, they must be reviewed by independent experts from the field in question. This scientific standard has been established over hundreds of years following the publication of the first scientific journal in Paris on January 5, 1665. The twelve-page, French-language *Journal des sçavans* reported on the latest literature and scientific findings. For example, the first issue contained an article on Descartes's *Treatise on Man*.

Naturally, peer review does not prevent all incorrect results, flawed methods, and false conclusions from being published, but it does protect against charlatans and has made modern research possible. It is crucial, though extremely slow.

It usually takes one year from submission to publication. If we aimed to publish an attribution study in the traditional scientific fashion, a year would pass before the public could find out whether a heat wave—with all ensuing deaths and crop failures—was caused by climate change and to what extent. By that point, the next summer would probably have come and gone, either cool and rainy or perhaps even hotter than the year before. Or there might be a hurricane on the other side of the world commanding everyone's attention. Whatever the case, very few people would still be interested in the high temperatures of the previous year.

Not only that, but a year would have passed in which our information could have proven useful. There is a huge difference between a one-off event and heat levels that will occur regularly thanks to climate change. If the event is likely to reoccur, it will be worth the effort for cities and villages to adapt to higher temperatures; for example, authorities can let people know where to go to cool off and can invest in public water dispensers.

The faster we are and the earlier we can make our answers public, the more impact we can have. Even in that San Francisco café, we suspected that Heidi's idea could prove a minor revolution.

Myles at least was not scared off. He had often graced the covers of *Nature* and *Science*—the most important scientific journals—with research that was different, provocative. I, on the other hand, had not yet realized that the concept of World Weather Attribution was far more than just a new, exciting project; but then I was still new to my job. While Myles had been devoting himself to other questions recently, I (as a research assistant on his team) had spent 2012 and 2013 testing and refining event attribution methods for various events and global regions.

I spent most of my time detecting and highlighting uncertainties in methods and results. Initially, this doesn't sound like something that would push our new discipline forward. Uncertainties occur not just in weather data, but particularly in climate models, both because they are a simplified version of the real climate system and because we can simulate only a limited number of weather events. The fewer simulations we have at our disposal, the greater the uncertainties. And the rarer the event, the fewer simulations there are—but these are the events in which we are interested.

Stating uncertainties is crucial, even if they make scientific results seem unnecessarily complicated. Take this made-up example: A study concludes that people who drink at least four cups of coffee a day live two years longer than those who don't drink coffee. Sounds good at first. But if the study had just a few participants, this average value of two extra years could mean that one half of the coffee drinkers lived three years less and the other half lived seven years more. It could also mean that all coffee drinkers lived longer: one, two, or three years more. Both scenarios have the same average value, but the findings are totally different. In the first case, the trend shows that drinking lots of coffee may or may not increase your life span; the second case shows that it will extend your life by at least one year. On the basis of this second result, the authors of the study would recommend drinking more coffee; on the basis of the first result, they would instead recommend conducting new studies with more participants to determine whether there is any possibility that coffee could shorten one's life span. Calculating and quantifying uncertainties are therefore crucial to every scientific study. To apply this treatment of uncertainty to our new methods, we first had to develop a standard procedure. Our initial studies were a little clumsy, but we improved every time.

And our team grew. At first, Myles and I were among just two small groups of scientists focused on attributing weather events. These included established climate researchers and newcomers to the field such as myself and (for example) Andrew King from Melbourne. A small, highly dedicated group who were always more vocal than most other climate scientists. Since then our group has grown significantly, but we still hold unofficial meetings once a year away from the major conferences to talk shop, engage in friendly arguments, and develop ideas for new methods. There is no agenda or final report, just twenty to thirty scientists bickering over the best statistical methods, framings, and approaches. We also discuss how many of our results we should present to the public—and above all, when. Although our core World Weather Attribution team focuses on attribution and believes it to be important, we are outsiders because we share our results with the public as soon as they are calculated.

Myles is no longer part of our team, but that's not unusual for him. When I bumped into him in the stairwell back in Oxford shortly after our meeting with Heidi, he told me that he had no intention of joining the project, much less running it. Instead, he assumed that I would take on the role.

If I did, I would be jumping in at the deep end. But when I thought about it, I knew the team would still have plenty of stubborn people. And for someone who had only just entered the world of science, this was a huge opportunity for me to determine what would be researched, when, and how.

Our founding team spent the next six months drawing up budgets, raising funds, and hiring postdocs to test climate models and analyze the data, all the while continuing our regular research projects. We also assembled a scientific advisory board of high-caliber climate researchers who would evaluate our work once a year and produce a report. This was imperative; circumventing the peer-review process meant that we needed another form of external assessment for our results. If the scientific community were to dismiss us as "unscientific," we would lose all public credibility and risk damaging the reputation of climate science, already under enough attack.

We developed most of the methods on which our results are based before the World Weather Attribution team was set up. They have been thoroughly scrutinized in numerous publications via the peer-review process, but our findings on individual weather events have not.

The Paris Hilton Event

After the first attribution analysis of a weather event in 2004, it would take seven years for another event to hit the headlines and be partly attributed to climate change. The studies caused a sensation not just within science, but also in the public sphere. Unfortunately, the headlines weren't exactly what you'd want.

In July 2010, a heat wave spread across western Russia. The people of Moscow suffered in temperatures of up to 40°C (104°F), and fires raged in the surrounding forests and heath. Hundreds of people died as a result of the heat. Scientists from Boulder, Colorado, who conducted an attribution study published in 2011 concluded that the heat wave was "mainly due to natural internal atmospheric variability."[2] In another study, however, scientists from Potsdam, Germany, calculated an 80 percent probability that these record temperatures would not have occurred without climate change. The media picked up on the apparent conflict between the results.[3]

Had our fledgling discipline proved itself useless already?

In the weeks that followed, I examined this question and used the method we had developed in Oxford to establish whether the studies really did contradict one another—and if so, which one was correct. The result was a pleasant surprise for all involved. Both studies were correct; they had simply asked different questions. One study focused on the heat record itself—the magnitude of the heat wave—while the other looked at the probability of the heat record being broken.

So had climate change made the heat wave more likely? The answer was a resounding "yes." Meanwhile, the actual temperatures were determined largely by local weather conditions and therefore had many natural causes too.

The difference in the questions posed might seem like a relatively unimportant detail, but it had a huge impact on the result. This experience taught us to look closely and ask precise questions.

Our study on the Russian heat wave (the third such study) was followed by several more, leading it to be dubbed a "Paris Hilton event" in professional circles: famous, but nobody quite knows why. Admittedly, this was fair neither to Paris Hilton nor to the victims of the heat wave.

Together, however, these various studies did achieve one thing: they established the attribution of extreme weather events as an independent, reputable, though small, branch of climate science.

Official recognition came in late 2012, when the prestigious *Bulletin of the American Meteorological Society* published a special issue featuring six attribution studies on events of the previous year. From then on, a special issue would be published every year featuring a growing number of authors and studies examining extreme events all around the world. In 2012, two of the six studies focused on Great Britain's previous warm winter.

By the time we met Heidi in San Francisco in late 2014 and decided that we wanted to determine the role of climate change in extreme weather while the world was still listening, we were more than a little aware that British winters were growing warmer; we had published a subsection in the fifth IPCC report, the epitome of peer review, released the previous year.

While IPCC authors are themselves scientists, the reports do not feature their own research or even the latest research; as noted, research work is included only if it has already been published in specialist journals and if the results and methodological quality have been confirmed by multiple studies. The reports are appraised both by scientists and by representatives of all international governments for maximum quality control.

Nevertheless, a number of scientists called us naive and even accused us of hubris when we initiated our plan to provide watertight studies on the role of climate change in global extreme weather—and to do so within just a few days, with little apparent methodical groundwork, while bypassing the tried-and-tested peer-review process, and with just a handful of researchers. It was as though, two years after the light bulb was first developed, we had announced that we were going to install electric lanterns on every street without knowing whether mass production was even possible.

So it's not all that surprising that some of our colleagues questioned our sanity.

The First and Second Cases: Europe, 2015

By the early summer of 2015 we were finally ready to start, or at least we thought we were. Now all we needed was an extreme weather event. The intense heat wave that had ravaged India back in May, claiming many lives, had come too early for us. We needed to start with something simpler and practice completing a study within a few days.

We didn't have long to wait: in July, temperatures far exceeding 30°C (86°F) swept across western Europe. The European heat wave of 2015 became our first case. Even in Oxford, it was so warm that arguments broke out over the few fans available in our institute and schedules had to be drawn up to make sure every office got a turn.

At least that's what I was told. Heidi and I were attending a conference in Paris. Although we didn't have a perfect situation, being away from Oxford, we didn't want to let this opportunity slip through our fingers.

Paris was sweltering and the heat wave had not yet abated. The forecast was our main source, rather than complete observed temperature data. Using this as our basis, we selected individual cities (rather than an entire region) that were expected to experience unusually high temperatures. We chose Mannheim, Paris, Madrid, Zurich, and De Bilt, a town near Utrecht that, as the headquarters of the Royal Netherlands Meteorological Institute, is superbly equipped with weather data.

People start to suffer when heat lasts for three days or more, which is why we decided to concentrate on the three hottest days. My colleagues and I spent the whole of July 1 and 2 in our hotel rooms, logged onto the Oxford servers, running calculations. We emailed back and forth during the night to reassure ourselves that we had analyzed the data correctly. We received results on the second night and wrote a fact sheet. After a sleepless night and two arduous, sweaty days in the hotel—with the occasional dispute over how to proceed and which method to use—we finally had our first real-time attribution study!

For the first time, scientific facts were available on the role of climate change in a weather event—and they'd been developed during the event itself. The next morning, we published a press release and the fact sheet, explaining that climate change had made the heat wave in De Bilt around twice as likely and a whole six times as likely in Madrid. The results for the other cities were somewhere between the two. Naturally, we hoped our press release would have an explosive impact.

But when we visited the websites of *The Guardian*, the BBC, and the *New York Times* the next day, there was nothing. Only a few niche media outlets had picked up on our results.

In hindsight, this was a good thing. It later transpired that Mannheim had reached 38°C, one degree higher than forecast, meaning that we had delivered results for an event that had actually taken a different form. Back in Oxford, we ran the calculations again and determined that the probability of Mannheim's heat wave had increased by 500 percent, compared to "just" 200 percent in the original study. If the first study had attracted major attention, it might have been very difficult to explain this to the public.

This experience taught us to wait until the event is actually over, at least if it develops quickly like a heat wave, rainfall, or hurricane. We also learned that our work is much easier if we write a scientific article at the same time precisely describing our method with all recordings and details—even if we publish it directly ourselves, rather than submitting it to a journal for evaluation. It's a pretty boring read but indispensable when repeating an analysis or reconstructing a colleague's approach to see how they achieved their results. We also learned that it's incredibly helpful for the team members to like each other and overlook each other's nighttime moods.

After this initial practical trial, by the fall we were ready for the next study. On December 5, 2015, Storm Desmond swept across the north of Britain and caused serious flooding. This became our second case. This time, it took us five days. We wrote a scientific paper, an easy-to-read summary, and a press release. On December 10, we held a press conference at the global COP21 climate conference in Paris and announced that, on average, the rainfall associated with the storm had been made 40 percent more likely due to climate change. This time the British press reported our findings, and did so with astonishing accuracy. Even outlets like the *Daily Mail* (not exactly amenable to climate science on the whole) reported the results of our study extensively and correctly. From our perspective, it was the perfect study. However, it caused quite a stir in the scientific community;

our colleagues were not convinced that our methods were indisputable.* To be accepted in professional circles, we therefore had to repeat the study and publish it in a specialist journal. This we did, with the results published in 2017.[4]

We had shown we could quantify the role of climate change in extreme weather events within a short time frame, even if we had to slow things down to convince our colleagues of the merit of the approach. It quickly became clear that our new approach had taken on a life of its own and could not be stopped. Attribution studies were now appearing in Japan and the U.S. shortly after extreme events occurred. However, definitive confirmation of our research integrity would come from an unexpected source.

The Climate SWAT Team

In light of the unease in the scientific community, the U.S. National Academy of Sciences (NAS) commissioned a report in late 2015 on the status of research into extreme weather attribution. The aim was to clarify whether events can be attributed and, if so, what types of events—and which methods are suitable. We were a bit apprehensive when our entire World Weather Attribution team was invited to

---

* Normally, articles have two or three peer reviewers; ours ended up having seven, because it was so controversial. While some of the reviews were very positive and others offered constructive criticism, two of the reviewers rejected the article as unpublishable because of the speed of the analysis. One year later we repeated all the analyses with further observation data and model simulations. The result was the same, and this time the reviewers were satisfied.

Washington, D.C., along with a few other leading scientists from the field.

In October, we found ourselves being questioned in an imposing NAS building on the corner of Fifth and East Streets. At least that's how it felt, even though the two days were organized like a scientific conference. It fell to me to give the first presentation explaining the status of attribution science. This was followed by discussions and presentations by specialists on specific categories of extreme weather. My job was to convince the authors of the report that our science was just as good as everything else happening within climate research. The audience remained largely silent, so I couldn't tell whether I'd succeeded. Nobody from our team was allowed to help write the report or assist with the review process. While this ensured that the report was truly independent, it did make the process feel a little like a tribunal.

We waited almost six months for the report to be compiled. In the meantime we worked on other projects and left World Weather Attribution on the back burner. After another interview via videoconference, the report was published in March 2016. It not only stated that our work can be applied legitimately to various forms of extreme weather, but even recommended that scientists use the method that our team had developed and refined.[5]

We weren't surprised; we had exercised even more caution than in our other research projects for sheer fear of making mistakes while under public scrutiny. Nevertheless, we felt we had achieved a victory—a victory sealed by an extensive article in the *New York Times* describing us as a "Climate SWAT Team."[6]

Of course, the report didn't silence all our critics within climate science,* but life would be pretty boring without a bit of opposition.

Now nothing stood in the way of a rapid attribution program. We could finally channel our energy into the research we actually wanted to do. Europe-wide heat waves, extreme rain in Paris, and storms in Great Britain are interesting, of course, but if the Seine starts to burst its banks more frequently, Paris is affluent enough to adapt. However, if climate change makes flooding more likely in Bangladesh, extreme weather can fast become a catastrophe.

Ultimately, we want to focus on extreme weather that robs people of their homes and causes injury or even death, extreme weather that can bring a country's economy to its knees or set it back years: droughts, heat waves of over 50°C (122°F), typhoons, and of course hurricanes. Developing countries feel the effects more than industrialized nations because they are more vulnerable, and extreme weather can threaten their very existence. Even within industrialized countries, marginalized regions and groups are affected more than any other.

Even where climate change plays no role, it is just as important—possibly more so—that interpreting an event not be left to lobbyists. Instead, we need to tread on the toes of

---

* One of our most vocal critics remains Kevin Trenberth, an eminent climate scientist from Boulder, Colorado. He argues that because climate models are not always capable of simulating the dynamic effect (the change in atmospheric circulation), we *underestimate* the impact of climate change. See also Lusk, "The Social Utility of Event Attribution."

those that failed to plan appropriately and ensure they take responsibility.

This is why we are fighting to get this scientific discipline up and running, fighting against a lack of weather data, unsuitable model simulations, and datasets that absolutely refuse to fit together. This takes time, particularly for extreme weather that we have not yet been able to attribute—for example, hurricanes.

Days 5 and 6

On Tuesday morning, President Trump dons a rain jacket and climbs up onto a truck parked in front of a fire station in Corpus Christi. Together with Rockport, this Texan city was the first to be hit by Hurricane Harvey. "What a crowd, what a turnout!" he shouts. Someone hands him a Texas state flag and he doesn't hesitate to wave it.

According to the *New Yorker*, at a subsequent briefing at a control center in Austin, the U.S. president comments that "nobody has ever seen this much water... The water has never been seen like this, to this, to the extent. And it's, uh, maybe someday going to disappear. We keep waiting!"[7]

Trump might have been even more surprised had he visited Houston, but he and his retinue decided to avoid the city—largely underwater, it is not the right setting for a presidential visit.

On Wednesday, the rain in Houston gradually stops. The extent of the damage is far from clear. Emergency workers in the area report that streets and houses are still flooded. One photo provokes an outcry on social media. It shows nursing home residents sitting in their wheelchairs, waist deep in murky water.[8] They are evacuated after the photo is published, just like thousands of other Houston residents who have been forced to lay out sleeping bags in schools and gyms.

But not everyone can be helped. No amount of preparation would prevent disaster relief services from being overwhelmed by one of the greatest floods in U.S. history. At least twenty-two people have died thus far throughout the district, almost a third of which is underwater.

Our team now knows that the city received its heaviest rain from August 26 to 28. The rain continued after this date, but the highest water levels were reached almost everywhere on August 28. In these three days, 41.07 inches of rain fell at the weather station with the most intense rainfall (Houston NWSO). These figures provide us with the information we need to begin our analysis and define the event.

We have determined the time frame, the three days with the heaviest rainfall. Next we need to narrow down Harvey's spatial dimensions. We try out various definitions: Houston alone, then Houston and its surrounding area, then as far as the state border—including Corpus Christi, where Trump stood and waved his flag, utterly surprised by the sheer volume of water.

At the very least, we can say that this is a truly exceptional event. Whether we focus just on Houston or look at the entire region, the weather station and satellite data return the same result: Harvey is an extreme event.

In our current climate, the quantity of rain that fell in the city should be expected no more than every nine thousand years—this, at least, is the conclusion of the largest, most complete dataset we have. But how can we use one hundred years of weather data to determine that an event can be expected every nine thousand years? With extreme value statistics. We take all observed weather events, whether extreme or not, and use statistical models to extrapolate events that are more extreme than anything we have seen before. The further we move from rainfall that has already been measured, the less precise these methods become, so these nine thousand years are just a rough estimate. The figure could just as easily be twenty thousand years, but not five hundred or fifty thousand.

In the coming days, we will receive new weather data and be able to provide more exact figures, although experience suggests that not much will change.

Now we need to determine whether climate change has played a role in Hurricane Harvey, and to what extent. While Harvey is certainly extreme, it may not be an "epic" and "historic" storm that came from nowhere, as Trump seems to have suggested. It is possible that the people in charge in Washington, D.C., Austin, and Houston should have expected more storms and rainfall of greater intensity than previously observed—that they could have known this would happen.

CHAPTER 4

THE HUMAN FACTOR

Calculating the Influence of Climate Change on the Weather

W HAT WOULD THE world look like today if climate change had never happened? This might sound like the concept for a utopian novel, but it is, in fact, the basis for our work. Only by simulating a world without climate change can we determine how it influences the weather of today.

To do this, we compare the weather that can occur in a world without climate change with the weather possible in our modern world. It's like taking a template of the weather possible in one world, placing it over a template of another world, and looking for bits that don't match—that is, whether the weather has become more or perhaps less extreme.

The starting point—the world untouched by climate change—is a purely hypothetical world that has never existed. After all, we are not asking how likely a weather event would have been in a world without humans. We are asking how likely a weather event would have been in a world like ours but without human-caused climate change.

In a world without humans, the atmosphere would be free from all the additional greenhouse gases we have pumped into it over the centuries. The vegetation would also be totally different; most of the Earth's surface would be covered in forests or in jungle very different from what we see today. For thousands of years, we have chopped down and reforested trees and, by traveling widely, have helped tree species spread from one part of the world to another. But the larger a forest is, the more it influences the climate and, by extension, the weather.

A world without climate change is therefore not the same as the world that existed centuries ago. Our model is no untouched, primeval world; it is a world of humans, just without additional greenhouse gases.

Our hypothetical world remains in the Anthropocene era, so it's a world in which humanity has at least made its mark—"Anthropocene lite," if you will. In a simulated world like this, the Industrial Revolution based largely on the burning of fossil fuels never took place. This is, of course, unrealistic; our planet would have developed differently without the Industrial Revolution—whether for better or worse is anyone's guess. But at least we wouldn't be grappling with the issue of global warming. And that is the sole focus of our counterfactual model.

Extreme weather attribution is based on the concept of simulating possible weather in this fictional world and comparing it with possible weather in the real world. The definition of an event (like a drought or hurricane) in the real world acts as a blueprint. We then search both worlds to determine how many weather events fit this pattern.

What precisely does this mean? First, we ascertain what weather is possible in a particular region under current conditions—for example, how much rain does Houston get on average? How intense is the extreme rain that falls there every ten years? Every fifty years? Every hundred years? How likely is the torrential rain we have observed?

We then ask the exact same questions again, this time simulating possible weather for a world without climate change. If a weather event is more (or less) likely in one scenario than in the other, then the difference is clearly due to climate change; this is the only difference between the two worlds for which we have run the weather simulation. If, for example, an extreme event occurs every ten years on average in our current climate but only every hundred years in the world without climate change, then climate change has made the event ten times as likely.

Measuring the World

First we need to reconstruct the weather in the real world, which will allow us to simulate potential weather in the modern world and then in the fictional world. To do this, we need to know about the weather occurring in the real world—that is, we need as many weather observations as possible dating back as far as possible, including temperatures, wind strength, rainfall amounts, and more.

On the one hand, we need these data to know what has actually happened: When and where did it rain, and how much, before floods occurred like those in Houston?

On the other hand, we need weather data to calculate the probability of the weather event in question. These data then serve as the basis for defining the event; for the Russian heat wave of 2010, we used the highest July average temperatures throughout a large area around Moscow (35° to 55° east longitude and 42° to 60° north latitude). Once we have clearly outlined the event, we can simulate it in the climate models—once in the modern world and once in the hypothetical world without climate change.

We also use weather data to check our models. Not all models can realistically simulate the weather events in which we are interested. In short, weather data are the link between climate models and the real world.

Without weather data, we would have only a vague idea of our weather's characteristics. Without many years of weather data, nobody would know that Kiel gets 27.5 inches of rain per year on average or how long the high-pressure areas that cause heat in the summer and cold in the winter persist in a particular region. Observed weather data are indispensable and precious.

Satellites Survey Our Weather

We have only been able to measure global weather since 1979, when satellites began to circle the Earth and record its weather comprehensively—just long enough to allow us to do our work today, at least in regions with no weather stations. Even if we had only thirty years of data, rather than forty, we could make statements about the climate today; in principle, the climate of a particular region is nothing more than the average weather over a period of thirty years.

However, thirty years is not long enough to say anything about changes in the climate.

This is no arbitrary value. Weather varies from day to day, month to month, and year to year—but also in longer cycles of ten to thirty years. Relatively little changes beyond this time frame, and the quality of data obtained over these longer timescales differs significantly. A thirty-year time frame incorporates the key spectrums of internal natural weather variability. There are, of course, changes on a scale of thousands of years driven by natural variability outside the climate system (e.g., the distance between the sun and the Earth).

The same would be true of a twenty-nine-year or thirty-three-year period. Ten years would definitely be too short, particularly in tropical regions where the weather barely differs from day to day but often changes drastically every five to seven years. We need to observe several such cycles in the average weather to make statements about the climate.

So thirty years is the minimum. We have to extend this time frame for extreme weather events, which are by definition rare and do not occur very often in a thirty-year period. Ideally, we would have records dating much further back. Some go back to the end of the nineteenth century, when climate change could barely be measured in the average global temperature, or even back to the eighteenth century; these weather observations are the best observed reflection of a world in which climate change never happened.

Today, living in a measured world is a matter of course for most people—geography has been measured at least since Google Earth was developed—and this includes the weather.

But even in the age of satellite observation, it is very difficult to tell whether rainforests in the central Congo are currently experiencing heavy rainfall.

At the moment, we have two types of satellites observing our weather. Geostationary satellites remain in sync with the Earth's rotation to provide images of one area around the clock, while other satellites orbit counter to the Earth's rotation and examine every point on the planet about every twelve hours. In contrast to geostationary satellites, these offer good spatial resolution but provide images of a particular region only twice a day. Some bouts of intense rainfall don't even make it onto their radar.

Every nation (or group of nations) with a space agency—like the U.S., China, and Europe—sends new satellites into space almost every year. They don't all focus on the weather, though; this technology can be used to observe forest fires, illegal rainforest clearing, and even munitions factories.

While President Trump and his government may be systematically cutting funds for weather and climate research and for satellite missions, NASA did manage to send one new GOES-R geostationary satellite into the Earth's orbit in early 2018 that should be able to better decipher what happens during a storm than was previously possible.

Satellites are particularly good at recording cloud coverage and large-scale areas of high and low pressure. They find rainfall much trickier—the old satellites can barely tell the difference between clouds that carry rain and those that don't.

At any rate, satellite measurements must be calibrated using other weather recordings and compared with data from a dense network of weather stations. This allows us to calculate satellite errors and make sure satellites provide the most useful data in the future. Ideally, satellites will at some point be able to image weather fairly precisely without the aid of weather stations.

Weather Stations Around the World

Weather stations are nothing special, really. Their equipment includes a thermometer, and the simplest use a beaker to measure rain volume. They usually have an anemometer to record wind speed, a barometer, and devices to measure relative humidity and the duration of sunshine.

Today, these instruments work more or less automatically in most weather stations, removing the need for manual daily readings. But even the best automatic stations need some maintenance. If a device malfunctions and is not replaced immediately, the series of measurements will be interrupted and become less useful—it may even be rendered worthless. Individual weather stations that our institute has set up in the Sahara, for example, frequently malfunction because the technology has failed or an animal has chewed at the cables.

To this day, the best and most complete series of measurements come from manned weather stations; for a hundred years, in times of war and peace, rotating teams have been reading devices at the same time every day, including national holidays. These include the weather station on the Telegrafenberg, a hill in Potsdam, Germany. It is now the headquarters of the Potsdam Institute for Climate

Impact Research (PIK), where I worked before moving to Oxford.

The data are immediately recorded in digital format and are technically available in real time. However, most countries do not release these data until days or weeks later—if they release the data at all.

The network of weather stations around the globe remains patchy and uneven—not only in the Sahara and countries like the Congo, where satellite technology is the only thing that keeps us from being completely in the dark, but also in Europe. The Netherlands may be densely packed with weather stations, but they are much sparser in France and Germany.

Logbooks: Learning From Seafarers

If we want to understand how weather is changing and why, we need more than just weather data recorded in the past few decades with the latest technology. We need to travel much further back in time, descend into the stacks of old European universities, and turn our attention to the people who sailed the oceans centuries ago.

For the seafarers of old, observing and recording the weather in detail was crucial to their survival, making ships' logs a unique source of weather data. The people behind the "Old Weather" project have set themselves the task of digitizing logbooks with the assistance of volunteers, citizen scientists who input weather data taken largely from the logbooks of whaling ships.[1] This is laborious work—most logbooks were written by hand and there is no way of telling at the outset which ones actually contain weather data—but this is what makes it so fascinating for many volunteers. Not only do they learn what the weather was like over the Arctic Ocean in a particular year, they also gain insights into life on ships that spent weeks at the mercy of the elements completely cut off from civilization.

In recent years, projects like this have enabled scientists to significantly improve time series for temperature, rainfall, and atmospheric pressure from the mid-nineteenth century onward. For the first time, weather data from the early days of industrialization can be analyzed—a gold mine for our work.

The Radcliffe Observatory in Oxford

Logbooks may be the most exotic source of old weather data, but they are certainly not the only source. The world's longest continuous rainfall records are kept three stories below my office in Oxford, in the basement of a brick building built in the 1960s.

These records come from the weather station at the Radcliffe Observatory in Oxford.[2] As is the case all over the world, this historic station housed in a tower has long since been replaced by shiny instruments embedded in a stretch of grass—at the foot of said tower in the grounds of Green Templeton College. The original documents containing the station's measurements are now stored in a steel cabinet in the basement of our institute. When I open one of the notebooks and skim the entries from the eighteenth and nineteenth centuries, I never fail to be impressed—less by the history they contain than by the fact that they have exactly the same kind of data as the twentieth-century entries (apart from the date, of course). This is what makes them so valuable: weather must always be measured in the same way if we are to determine whether and how it is changing.

All around the world, people are entering old weather records into computers and making them available to the scientific community. All the weather data from the Radcliffe Observatory have now been digitized, and I no longer need to go down to the basement.

We used these data to investigate the extreme rainfall in southern England in 2014, examining the volume of winter rain in Oxford over the last two hundred years.[3] Based on these records, we found that extreme rainfall like that of 2014 has been made around 40 percent more likely due to climate change. Similarly, we used the Radcliffe temperature data to study the heat wave in the last week of July 2019. The historical high temperature record of 35.1°C (95.2°F) set in 1932 was broken by 1.4°C, setting a new record of 36.5°C (97.7°F). In a rapid attribution study, we found that climate change made a July heat wave as observed in 2019 about five times as likely in Oxford.[4]

Computer Simulations: Rolling the Weather Dice

However, even the best observation data reflect only the weather that has actually occurred. The data don't reflect all possible weather scenarios.

But how can we talk meaningfully about an event to be expected every ten thousand years when we only have observation data from the last hundred years or so? We can't. It's as simple as that. That would be like rolling a die ten times, getting a six five times and using this information alone to calculate the probability of rolling a six.

So what do we do?

If we know the distribution of the weather data, we can effectively extend the observed data. We may not know exactly how the data are distributed, but we can at least make justified assumptions and use them to gauge the likelihood of the event.

And so we need statistics. Whether extreme or normal, our daily weather is only ever one of the many possibilities within our climatic conditions. To determine the likelihood of an event, we need to consider both the actual and possible weather. If we want to calculate the probability of a six solely by reading the number on the die, we need to roll it more frequently. In principle, this is exactly what we do with our climate models: we roll the weather dice. To simulate possible weather under the given climatic conditions, we need statistical models and climate models.

Now it gets even more complicated. Our next step is to leave the world in which we live and enter a world without climate change. This means we have nothing with which to compare our models. This is where attribution actually begins. To gauge the likelihood of a weather event in a world without climate change, we need to measure possible weather under climatic conditions we have never observed. While we know for certain that we have a one in six chance of rolling a six, we don't know at the outset what weather is possible in an atmosphere that hasn't been manipulated.

To find out which weather is possible, it is not enough to simply use a climate model and simulate possible weather once or twice on every single day in the past ten years. Instead, we need to simulate possible weather several hundred times. For example, simulating just two possible summers would be like picking two clovers, one with three leaves and one with four. If you had never seen a clover before, you wouldn't know that three-leaf clovers are normal and four-leaf clovers are the extreme event. This shows why we need to simulate possible weather so many times.

These massive simulations would be impossible without the rapid development of computer technology with more efficient processors and larger memories. We simply couldn't afford to calculate such "ensembles" of model simulations. The British climate portal Carbon Brief has calculated that "a global climate model typically contains enough computer code to fill 18,000 pages of printed text... and it can require a supercomputer the size of a tennis court to run."[5]

In fact, we still can't afford the enormous processing power required for our attribution studies. Essentially, we are only able to do what we do because of a very special, adventurous community of people who hunt for aliens in the vast expanse of space. I'll come back to them later.

First, we need to immerse ourselves in two fictional worlds: a world with the potential weather of today and the world as it would have been without climate change. To enter these worlds, we need climate models—and physics.

A World Without Climate Change

A climate model is a mathematical representation of the climate system. We can use it to recreate the climate system and create a sort of artificial Earth in which we can run experiments—after all, medical students practice on dummies before operating on actual people.

Like all physical systems, the climate system is determined by laws of conservation for energy, mass, and momentum. In accordance with these laws, energy, mass, and momentum are neither created nor destroyed, but simply convert themselves into different forms within a closed system.

The Earth's climate system is not a closed energy system; its energy comes from outside, from the sun. But because its energy is not destroyed, equal amounts of energy must enter and leave the system. This is the first key equation on which every climate model is based: the law of energy conservation. A simple climate model of this type can be used to calculate how much the average global temperature changes when more greenhouse gases enter the atmosphere or a volcanic eruption adds more sulfur particles. While the average global temperature is crucial, we want to know more; when consulting with a patient, a physician has to do more than simply take their temperature.

The next, more complex stage of climate modeling also incorporates mass conservation. The Earth can be viewed as a closed system for mass; the mass of the particles that enter and leave the atmosphere is so low compared to the mass of the atmosphere itself that we can ignore it. The conservation of mass means that if the pressure in the ocean or atmosphere decreases in one region, thus altering the number of water or air molecules, the pressure must increase in another area of the system because molecules can't disappear.

Mass conservation can also be calculated using a simple equation. This can be used to divide the climate system into very large sections. The Earth is separated into boxes: one box for the tropical atmosphere, one for the temperate zone, one for the poles, and one for the oceans. Each box contains pressure and temperatures that change if, for example, more energy enters from outside. We can calculate how the differences in pressure between the boxes balance each other out—in other words, air circulation and oceanic currents on huge scales.

These (still very simple) models can be simulated quickly and often to calculate the different speeds at which the land, atmosphere, and oceans are warming up. However, we still cannot use these models to simulate weather. For that we need the conservation of momentum, Newton's second law, which states that force is equal to mass times acceleration.

If you know the force to which air molecules are subjected at every point on Earth, you will know how these air molecules will move and, therefore, how the wind will blow. When they first derived what would later be known as the Navier-Stokes equation, physicists Claude-Louis Navier and George Gabriel Stokes described the four forces that act upon and accelerate air molecules: the Earth's rotation (Coriolis force), the differences in atmospheric pressure

(pressure-gradient force), the force of friction, and the force of gravity.*

We now have a complete general circulation model with which to simulate weather—at least in principle. We need to simplify these equations describing the conservation of energy, mass, and momentum in the climate system because models cannot solve all equations for every point in the atmosphere or ocean; that would require infinite computing time. Statistician George Box is considered the first to have said that all models are wrong but some are useful[6]—and he was right. While all climate models are simplified representations of the actual weather, they are certainly useful if they correctly reproduce key characteristics.

The greatest simplification in climate models is to consider the atmosphere not as a continuum, but as divided into noncontinuous three-dimensional pieces, forming a grid or network covering the globe. The simplified equations are solved at every grid point. In old climate models, the distance between grid points could be several hundred miles; in newer models, the distance is around 62 miles at the equator and less at the poles. Meanwhile, the vertical

---

* There is currently no general analytical solution to this equation—whether a clear solution exists is one of the great unsolved mathematical questions. However, we do not need a general solution that applies to every molecule, time, and place to simulate the weather using this equation. Instead, we are looking for a solution that is precise enough to describe the movement of the air fairly correctly but does not require so much calculation that it would be too expensive to apply to the entire atmosphere. Achieving this balance is not easy, and all climate computing centers worldwide experiment with various solutions in their models.

distance between grid points increases the higher they are in the atmosphere. This is because our weather takes place in the lower atmosphere and a lot happens in a small amount of space; farther up, the pressure is so low that there are very few molecules whirling around and changes play out on a larger scale. The greater the distance between horizontal and vertical grid points, the faster the models can calculate—but less realistically.

At every grid point, the model calculates the temperature, pressure, wind speed, and wind direction (along with many other meteorological variables) based on the prescribed time frame—for example, every fifteen or thirty minutes. To make this happen, we need to feed the equations in the model with initial values for all variables before starting the model. Instead of setting these values to zero, which would require immense computing power, we provide the model with initial values for variables such as temperature and wind speed so that it can calculate the changes. These are based on values that have been observed. In addition to these initial conditions, the model still needs to know what factors are driving the climate system outside its scope of calculation: current solar radiation, the concentration of greenhouse gases in the atmosphere, and the concentration of tiny dust particles ("aerosols")—what scientists call the forcing conditions. If the model's grid points were just a few feet apart, the model would now be complete: physical equations, initial conditions, forcing conditions.

However, the grid points are often tens of miles apart, yet weather occurs on a much smaller scale. We've all visited a city and found it to be raining at one end and sunny at the other. Therefore, this form of rough climate model cannot resolve rainfall. As a result, rainfall does not form part of the equations based on physical laws—just like cloud coverage and many other variables that exhibit small-scale changes.

There is, however, a solution: parameterizing the variables. This means that whether and how much it rains at a grid point is determined not by a physical equation, but by an empirical equation; we are looking for a link between the variables that the model can calculate by applying physics and the variables we need that cannot be calculated using physics. Models used to predict economic developments are based solely on such empirical equations. There is no law to prove that unemployment drops when the gross national product rises. Empirically, however, this is certainly what happens, which is why economists use current unemployment figures as a parameter to predict economic growth.

In principle, we follow exactly the same procedure when calculating rainfall in climate models—comparing climate models with weather records and testing which parameters we need to modify to better simulate the actual weather. For example, we need to test how large a drop of water needs to be in a cloud before it falls as rain. It is rare for parameters to realistically represent the weather all around the world. A parameter value will often simulate rainfall very realistically over Germany, but also turn the rainforest into a desert. We might be able to tolerate this if we only wanted to forecast European weather, but if we want to learn about changes to the global climate, then we need compromise parameters—parameters that generate fairly realistic weather for the entire world but do not necessarily fit in with the physics (for example, raindrops that are much larger than possible in the real world but generate realistic volumes of rain in the model).

Even without these simplifications, models would not be perfect because the climate system is chaotic. Even a minor change in the initial conditions might completely alter weather developments—the proverbial butterfly effect.

Despite these uncertainties, climate models can achieve a great deal. Global warming itself has shown that climate models work: forecasts from the early 1990s predicted the average global temperature from 1990 to the present very accurately considering the difference in actual and predicted greenhouse gas emissions—and despite the climate models of the time working with much rougher spatial resolutions.

What our models cannot do is precisely predict the weather in ten, twenty, or thirty years. The climate system is too chaotic for that. But it doesn't actually matter whether it will rain in Oxford on January 30, 2050. All we want to know is how likely it is that January 30, 2050, will see as much rain as in every average January of the last two hundred years. Has this probability changed, and if so, by how much? Our answers are constantly improving.

To answer this question, we need more than one climate model so that we can identify errors in individual climate simulations caused by the model parameters. The more models return the same result, the more confident we are that this result reflects reality. And the more simulations we run in a model, the better we can gauge the likelihood of a weather event. For a good attribution study, the models need to translate the physical equations and empirical equations into computer code in as many different ways as possible. This, however, requires enormous computing power, producing files of several terabytes, most in different formats.

Ultimately it's all about the numbers, lots and lots of numbers—so many that it can take up to two hours to import files containing one year of simulations from just one simple model. Unless we are studying something like heat waves that span a whole continent, we need models with a high spatial resolution—a dense grid—to calculate weather. We obtain these model simulations from the climate models of major meteorological computing centers like the European Centre for Medium-Range Weather Forecasts (ECMWF), the U.S. National Center for Atmospheric Research (NCAR), and the Japan Agency for Marine-Earth Science and Technology (JAMSTEC). It can take two weeks to simulate one model year. We would need plenty of patience and money to run these models on just one server. And we don't have either of those things.

Learning From Alien Hunters

This dilemma was solved by alien hunters, of all people. In the 1990s, UC Berkeley found itself with a problem: huge quantities of audio recordings from space that had been collected via radio telescopes and might contain proof of extraterrestrial life. Since nobody knows what this proof might sound like, the laboratory team couldn't let a machine listen to the recordings because they had no examples with which to teach it. Human labor was required. Since the scientific team was far too small to trawl through the masses of data, they started looking for helpers.

They developed a piece of software, known as BOINC, that sent audio recordings to private computers all around the world owned by people who volunteered to listen and report anything interesting they found. The SETI@home project was born, and it's still going on today.

We follow a similar pattern. We even use the same software to solve our modeling problem with the aid of thousands of volunteers around the world. There's just one difference: instead of donating their time and actively searching for aliens, our volunteers provide us with time on their computers and, essentially, money by paying slightly higher electricity bills. This gives us access to by far the largest server in the world.

Thanks to the dedicated volunteers who support the climate*prediction*.net project, we don't pay a cent to use this computing power. In 2015 alone, these computers racked up 120,000 years of processing time. Even the cheapest cloud service would charge us $6 billion for that.

Our program runs in the background on volunteers' computers. Anyone who essentially uses their computer like a typewriter—giving the processor very little to do—can lend us their processing power. If a computer reaches capacity, our model stops. Participants don't need any technical or scientific knowledge; all they do is download the BOINC software, link up with the climate*prediction*.net project, and let their computer calculate weather for a model year. When the results are ready, the file is sent back to us for analysis. For security reasons, we never have access to the volunteer's computer. We also offer a screen saver for most of our experiments and anyone interested can watch the calculations run and the changes in temperature, pressure, or rainfall.

Some of our around forty thousand volunteers keep hold of their old computers just so climate*prediction*.net can use them. Participants come and go over the years, some joining in when we run an experiment focusing on their region of the world. Over the project's life span, 700,000 people have volunteered. Without this project, attribution science for extreme events would probably have come to fruition several years later.

Day 7

On the last day of August, the rain stops. But the streets of Houston are still filled with water—toxic water mixed with oil, gasoline, corpses, and all manner of garbage. Nevertheless, the people who could not or did not want to flee the deluge wade through the water or use any type of vessel they can find. They have no choice; they need to find food and someone to help with the cleanup. These people are now the highest priority for the disaster relief teams.

The international media have shifted their attention once again. Having focused first on the hurricane itself and then on the people suffering its effects, they now report on the broader impact and ask whether the event could have been predicted. Did the city do everything it could?

They also raise the question we have been trying to answer since yesterday: What kind of phenomenon was this? Can we really expect a city to prepare for an event expected every nine thousand years or so? This figure has not yet been quoted, but we are now fairly certain that this is the approximate magnitude of the event's probability in the modern world (on a planet already 1°C/1.8°F warmer on average than in preindustrial times). To find suitable observation data, we don't need to go down to the basement or consult sailors' logbooks. Rainfall records for the U.S. are digitized and freely available. We can simply download the data and identify Houston using a number from the World Meteorological Organization.

We can already see from the weather data that climate change has, in fact, increased the probability of heavy rainfall like that in Houston—and to a greater extent than would be expected through global warming alone.

CHAPTER 5

HEAT WAVES, DOWNPOURS, AND MORE

The Role of Climate Change in the Weather

FTER MORE THAN two dozen attribution studies, I am gradually developing a feel for the way that climate change manifests in our weather—and it probably isn't how most people imagine.

Previously, even climate scientists didn't know exactly how it worked in individual cases and had to stick to generalizations—for example, saying that global warming has increased the risk of heat waves, stronger cyclones, and heavy rainfall on a global average. However, the fact that climate change influences the weather and we know the average does not help us to prepare for the changes this will bring where we live. And it isn't as powerful as localized, immediately relevant evidence in influencing both the general public and politicians to make changes.

Our team offers a fairly exclusive insight into how the climate manifests in specific weather and how climate change has altered the risk of certain types of extreme weather. Our studies have allowed us to identify initial patterns.

This isn't easy. Climate change doesn't work like a stimulant that systematically spreads over all the world's weather events and fuels them equally worldwide—at least not in all the places where the circulation is also changing and feedback processes are at play.

The effect of climate change on our weather is extremely erratic, almost petulant: it might make an event more likely, make it less likely, or have no effect on its frequency whatsoever. In all cases, climate change does influence the weather, but to different extents and with very different consequences.

Heat Waves in Europe: The New Summer Norm

Climate change is reflected more strongly in heat waves than in any other phenomenon. When we began our work, we still believed that heat spells would be the easiest form of extreme weather to understand and simulate—the changes are great, requiring only relatively rough models and few simulations. We thought we knew more or less everything already, but we were wrong; more intense heat waves are an extremely complex sign of climate change.

In summer 2003, a heat wave lasting several weeks spread across Europe. Temperatures of over 40°C (104°F) were recorded in several German towns and cities, while southern Portugal hit 47.5°C (117.5°F). This was torture for many elderly people; the heat was too much for their hearts and circulatory systems and they collapsed in the streets or in their homes. French funeral parlors were so full that refrigerator trucks were repurposed for corpses.[1] French researchers estimate that 70,000 more people than usual died in Europe that summer.[2] It was one of the continent's most devastating natural disasters and made people realize that heat waves can kill, even in Europe.

When Europe's next big heat wave came in summer 2006, France was better prepared. The authorities advised the public to avoid staying out in the sun and to drink plenty of water. They also allowed people to access public buildings if their homes were getting too hot. Europe seemed to have a reasonable handle on the new heat conditions.

Perhaps the most surprising aspect of heat waves is something that sounds pretty mundane at first: climate change affects temperatures differently around the world, even within the same heat wave.

Here's an example. Meteorologically speaking, broad swaths of Europe were covered by a large, stationary high-pressure area in June 2017. Heat records were broken at London's Heathrow Airport and in the Bavarian town of Kitzingen. Huge forest fires spread across Portugal, kindled by the heat. Naturally, absolute temperatures varied from country to country: while the 37°C (98.6°F) achieved in Paris didn't break France's 2003 record, in Great Britain June 21, 2017, was the hottest day for more than forty years at 34.5°C (94.1°F)—a record broken again in 2019.

Of course, the extremity of a weather event also depends on whether a record is broken by 0.1°C or 1°C. Thanks to climate change, Belgium can now expect a heat wave like 2017[*] every ten years, Spain only every eighty years. Accordingly, climate change has made this heat at least four times as likely in Belgium and increased its probability by at least one order of magnitude in Spain. Previously, a heat wave like this was almost outside the range that Spain could expect; suddenly, however, it has become a real possibility, if an extreme one. In Belgium, meanwhile, the formerly extreme event of 18.1°C (64.6°F) as the mean June temperature has become the new norm.

Unpacking such differences won't make the headlines and may appear highly academic, but it is these differences that turn a pleasant summer into a costly natural disaster. And if climate change drastically alters the probability of such a natural disaster, then countries need to revise their plans.

Our studies have revealed how vulnerable climate change has made us. However, societies in other parts of the world are far less prepared for heat waves. In 2015, the Indian state of Andhra Pradesh experienced sweltering heat of up to 48°C (118°F); more than 1,800 people died, particularly in the slums, which are neither air conditioned nor shaded by trees.[3] We established that a heat wave like this had become around twice as likely due to climate change.

---

[*] In this case defined based on the average daily temperature, as this was particularly extreme.

Extreme Rainfall

The world is getting hotter. This means not only a greater risk of heat waves worldwide—differing according to region—but also more rain on average. A warmer atmosphere can hold more water vapor. If the Earth becomes 1°C warmer, we expect rainfall to intensify by roughly 7 percent. This connection was discovered by Rudolf Clausius and Benoît Paul Émile Clapeyron in the second half of the nineteenth century.

What interests us, however, is what this means for specific cases. Do other factors also come into play?

On December 6, 2015, it rained in Great Britain. Not exactly out of the ordinary, except that it rained a lot. And I mean a lot. Storm Desmond was sweeping the nation. We discovered that climate change had made extreme rainfall between 5 and 80 percent more likely. This margin of uncertainty might seem relatively large, but it isn't: the lower limit is greater than zero, which means that climate change has definitely made this event more likely. The upper limit indicates that it hasn't become twice as likely; otherwise it would be 100 percent. If an event like Desmond could once have been expected approximately every hundred years, now it will occur roughly every seventy years—much more frequent but still rare enough that, statistically, a British citizen will only experience it once in their lifetime.

Climate change plays a role in many cases of extreme rainfall, although to a much smaller extent than for heat waves. Building your house on a floodplain puts you at much greater risk of a flooded basement than climate change.

Our studies show that similar regions and seasons will return very similar results for similar types of extreme rain (this may seem self-evident, but it isn't; the dynamic effect could have complicated things considerably). It seems almost superfluous to continue with individual studies.

Things get more interesting when we look at subtropical regions like Louisiana on the American east coast. In August 2016, the state was hit by the worst rainstorm in its history. Around 100,000 houses were damaged and thirteen people died in the floods.

Our attribution study showed that climate change had made the rainfall at least 10 percent more intense—more than the 7 percent calculated by Clausius and Clapeyron. This means that both global warming and altered circulation are playing a role. Global warming isn't the only phenomenon to make it rain.

Climate change is much more prominent in subtropical rainfall. In the case of Louisiana, we established that global warming had doubled the probability of this torrential rain, although our analyses did not rule out the possibility that climate change had made it up to ten times as likely. The scale of these results is more reminiscent of heat waves than of Europe's relatively tame winter rainfall.

A Lack of Cold Spells

Climate change also dramatically affects another phenomenon barely mentioned by the media, despite its grave consequences: the lack of cold spells. We only talk about the cold when it gets really chilly.

There should be far more discussion about the fact that climate change is making our winters ever milder and frost days increasingly rare, as seen in Great Britain in November 2011.

This was one of our first studies in Oxford. Our methods were still a little clumsy, but the result was clear: in our world with climate change, a November without night frost (like that of 2011) will occur around every 20 years. Without climate change, it would happen every 1,250 years.[4]

A lack of night frost in November sounds like a pretty boring nonevent, but a whole winter without any frost would have serious repercussions. More people would suffer bites in spring and summer because fewer insects would have died off than in a "normally" cold winter. There would also be more parasites to afflict livestock and crops, fruit and vegetables—and in their attempts to contain these parasites, farmers would use more pesticides on their fields and farmland. In addition, many crop plants have seeds that need a period of cold to break dormancy; so no frost, no plants.

Even regions of the U.S. dominated by climate change deniers are starting to see these effects—although it will probably take a few frost-free spells for people to truly realize that this is more than just bad luck, more than just natural variations in the weather.[5]

In winter 2017, the small American city of International Falls near the Canadian border was literally frozen. Temperatures of –38°C (–36°F) were recorded. Residents said it felt like their skin was burning.

President Trump commented on the cold snap from his vacation in Florida, where it was 24°C (75°F):

In the East, it could be the COLDEST New Year's Eve on record. Perhaps we could use a little bit of that good old Global Warming that our Country, but not other countries, was going to pay TRILLIONS OF DOLLARS to protect against. Bundle up! (@realDonaldTrump, December 28, 2017)

Our World Weather Attribution team analyzed this cold spell and another in southeast Europe in January 2017. In both cases, climate change had made the cold less likely. Without it, it would have been even colder in the same weather situation. In the face of global warming, these results aren't necessarily surprising.

There is, however, also a theory that cold spells are becoming more frequent, particularly in the U.S. According to this theory, disappearing marine ice in the Arctic is making the polar vortex weaker. The polar vortex is a weather system that remains mainly stationary over the North Pole in winter and separates polar air from other atmospheric circulation. At some point, this system collapses and the cold polar air heads south, spreading out over the continents.

Scientists can simulate this effect in climate models. However, the observation data do not indicate that this phenomenon is occurring more frequently or that it is strong enough to counteract the effect of global warming. This is one of the debates currently taking place within the scientific community, an area in which our understanding is developing and expanding. The outcome remains open. Personally, I would prepare for warmer winters.

When Climate Change Takes
Itself Out of the Equation

Sometimes climate change is involved in a weather event but neutralizes itself, concealing the part it plays.

Droughts are a good example of this. We could assume that aridity is decreasing around the world because extreme rainfall is increasing. But droughts are more than just a lack of rain—at least in regions with humid climates where more rain falls than can evaporate. Here, evaporation plays just as big a role as rainfall. So more rain is falling thanks to global warming alone, but more water is also evaporating. Droughts become more or less likely depending on which of these two effects is stronger.

However, both effects might be equally strong and the risk of droughts might not change at all. We determined that this was the case for a drought in the São Paulo region of Brazil in 2014. The study is relatively old, but the methods are excellent.[6] We conducted individual studies of the two most important meteorological factors in a drought: rain and evaporation. The likelihood of rain increased due to climate change, but evaporation also increased. If we combine these two variables and consider the actual risk of drought, then the two effects cancel each other out; this, at least, was the case in São Paulo in 2014. Climate change played a major role, but the drought risk had not changed.

Our study went one step further and analyzed water consumption too. In the years leading up to the drought, water consumption had risen exponentially. This meant that the drought had a much greater impact than ten years ago, regardless of climate change. In other words, what looks like a weather event fueled by climate change probably became a disaster due to lack of planning or the ill-advised use of resources.

The flooding of the River Elbe in 2013—or, to be more precise, the rainfall that led to major flooding of the Elbe and Upper Danube in May and June 2013—is another example of a weather event in which climate change played a role but did not affect the risk of the event occurring.[7] Although thermodynamics suggest an increase in such phenomena, the observation data (statistics) and model simulations (physics) both concluded that the probability had not changed. Therefore, dynamic changes in the frequency of low-pressure areas must have counteracted the thermodynamic signal.

The climate will of course continue to change as temperatures rise, and this balance may alter at some point. We did not consider this in our comparatively old study on the Elbe flood, but we did include it in our study of the São Paulo drought and all newer studies. We can project that this balance, in which climate change neutralizes itself, will continue for São Paulo even if the planet grows 2°C (3.6°F) warmer.

The Great Unknowns: Hailstorms, Tornadoes, and More

If we had perfect models and weather data for all the factors that drive various weather systems and events, then this chapter would end here. Unfortunately, we don't. We are still unable to provide reliable results for some types of extreme events. These include hailstorms, tornadoes (like the German tornado of May 2018), and also meteorological events that take place on such a localized scale that they cannot be simulated in regular climate models. We don't even have any really useful observation datasets for hailstorms that can tell us when, where, and how many hailstones come raining down.

For some other events, it is only as we conduct a study that we realize that none of the climate models can provide reliable simulations, and that prevents us from determining the extent to which climate change is involved in a specific event. Climate models still leave a lot to be desired, particularly in regions such as mountains where weather conditions can differ greatly on a small scale. It is equally difficult to simulate the circulation of monsoon rains at the right time and in the right place.

You won't read about these events in the general media or even in specialist literature. Events in which climate change can be shown to have played little or no role barely make the headlines. Studies never completed due to a lack of data and models find it even harder to attract attention; "we tried but failed" doesn't make a good story. While this is understandable, it skews public perception of climate change.

The media tend to pick up on studies in which climate change plays a clear and major role. This creates the impression that climate change is making everything worse. Sometimes that's true, sometimes it isn't—for example, the drought in Brazil. And sometimes it's very convenient to have big, bad climate change as the culprit.

It is just as difficult, if not more so, to communicate about events that have become more likely according to the observation data, but for which model simulations show no or very little change in probability. Sometimes this is because models do not simulate key processes realistically. If we spot this, we can weed out these models. In other cases, however, models pass all the tests despite contradicting the weather data—and then we have a problem.

When Humans Counteract Climate Change

There are also cases in which climate change is definitely involved but its effect on the weather is negated by humans themselves. This was the case in Phalodi in northwest India, where a record temperature of over 51°C (124°F) was recorded on May 19, 2016.[8] Together with colleagues from Delhi, we performed a study and established that the probability of such an event had not increased.

This result might seem surprising at first, given that mean temperatures have increased in India too, but this event is driven by many factors. First, a great deal more water is being used in the region today to irrigate fields and farmland, making the air cooler and moister.

The air is also highly polluted with particles of all varieties, which reflect the sunlight and cool the air. Climate models are very bad at reproducing the interplay between radiation and nanoparticles, making it difficult to test this theory. Nevertheless, it deserves serious consideration. If it transpired that these particles were masking the effect of climate change, and air pollution were to improve in the future, then maximum temperatures would increase massively at some point. There are other examples of this: maximum temperatures in Europe soared after industrial operations came to an abrupt halt in the former Soviet states in the early 1990s and the air became cleaner. Don't get me wrong, I'm not advocating dirty air; air pollution kills a lot more people than heat.

Results such as those for the Indian drought can also be interpreted as follows: climate change is not yet playing a major role because human activities are masking its effects. But climate change could soon be making a much clearer impact, and we need to be prepared.

Looking to the Future

Attribution studies can only ever capture one moment. While this is usually sufficient to isolate the effect of climate change, simulations of the near future are more informative in some cases (like the Indian heat wave). There, we can see that whatever phenomenon is currently counteracting the effects of climate change will eventually be thoroughly overshadowed by climate change itself.

Naturally, the methods we use to calculate how the weather has been altered by climate change to date can be applied to more than just the past. We can compare our real world both with a world without climate change and with the possible world of the future—allowing us to simulate the weather in scenarios in which the Earth is 1.5°C, 2°C, or even 3°C, 4°C, or 5°C (2.7°F to 9°F) warmer.

Projections of the future are also an important way to check the results of attribution. If we have established that climate change plays a clear role and climate models predict a similar, but stronger, effect for the future, this will boost trust in our statements. If the projections indicate very different trends, however, then we may have understood less than we think. Placing the results of our studies in the context of longer-term forecasts also increases potential uses of our studies. If we know that climate change plays a major role and extreme weather isn't just bad luck, then politicians and disaster relief services can prepare accordingly.

Day 15

On the morning of September 8, 2017, I sit in my office in Oxford and open an email from Geert Jan succinctly titled "Update on yesterday's call." It answers the question of how much climate change has to do with Harvey's rainfall.

The storm bears one of the clearest signs of climate change of all extreme rainfall we have studied to date. Climate change has made this type of torrential rain around three times as likely. This is the conclusion of the statistical comparison between the simulations of the current climate and of the world as it could have been without climate change.

In other words, this sort of downpour would be much rarer without climate change. With a probability of occurrence of once every nine thousand years, this event is still highly unlikely, but rainfall like that of Harvey would become three times as likely with every further 1°C (1.8°F) of global warming. At some point, an event that might be experienced by one generation in hundreds will need to be factored into long-term planning at least.

From the outset, it was clear to me that the result would be a range, not a specific number. The best estimate is that Harvey has become *at least* twice to four times as likely. We cannot rule out the possibility that it has become ten times as likely.

This isn't a totally new result; we calculated it using the weather data while defining Harvey. Today, however, we can compare it with Geert Jan and Karin's model simulation, which shows a slightly lower probability than the observation data, although within the range of uncertainty. The results are not identical, but they do overlap.

So far we only have *one* model. For the U.S. model, we only have the data from the Louisiana study of the previous year, which didn't include Houston. And it has taken an annoyingly long time for the model on the Mexican server to make its way to a server we can use for analysis. We will need a few more days to incorporate further models into the study. So this is more of an interim result, at least if your standards are as high as ours. In principle, however, the result is solid, partly because it closely resembles the study of the previous year.

The result didn't surprise me. Like the results of almost all our studies, it developed gradually, with no fanfare. But it's still satisfying when the pieces of the puzzle finally slot into place.

Now we need to decide whether to hold back or publish this result immediately. I argue that to publish immediately would be to ignore our self-imposed standards, given that we've only calculated one model so far.

Other team members point out that the model and the observation data match and that both reflect what we expect from the physics. There were no surprises. And the information we have gleaned has essentially already been publicly announced, if only in terms of trends and without specific figures. Now that we have the figures, the pressure is mounting to address the public before Harvey disappears completely from the media spotlight.

But this is exactly what seems likely to happen; a new hurricane has started to weave its path of destruction across the Atlantic. Everyone is talking about Hurricane Irma, which hit the Caribbean island of Barbuda on September 5 with wind speeds of 186 miles per hour, leaving indescribable carnage in its wake. Almost all buildings have been destroyed and all 1,800 residents were forced to relocate to the neighboring island of Antigua. Only dogs, cats, donkeys, and pigs remain, roaming the ruins, masterless and starving.[9]

Ideally, we'd now have an answer for Harvey *and* Irma. But that would require a much larger team of scientists focused on nothing but attribution studies.

This is a real dilemma, and I still haven't found a solution. Our fastest study was Great Britain's Storm Desmond, which took five days—five days in which we focused on that and nothing else. We won't beat this record until the 2018 heat wave in northern Europe, a less complex extreme event for which we have developed sufficient routines. To answer the attribution question for two hurricanes, we need either lots of time and people or, at the very least, lots of experience with the event and lots of people too. The problem can be solved in the longer term, but right now our situation is urgent and highly frustrating.

In the end, a colleague at UC Berkeley makes the decision for us. Climate scientist Michael Wehner tells us that he is also working on an attribution study, although his is based solely on observation data, so it's really more of a detection study. And he's just sent it to a scientific journal. This isn't a huge surprise; Michael and his team have been working on hurricanes for a long time, and he was convinced from the outset that real-time attribution studies are unbelievably important in communicating the significance of climate change—particularly in the U.S. From a scientific perspective, a second study is amazing; more data, more methods, and independent scientists will boost confidence in our results. But from a communication perspective, it isn't necessarily good news. Different methods mean different figures, particularly since the second study does not include attribution models. Even the slightest of differences between the studies will make communication more difficult.

If you analyze just the observation data, you can determine whether the probability of an event has changed, but you can't say whether this change is driven by climate change or other factors. This isn't attribution. Michael isn't claiming any differently, but this crucial distinction won't necessarily be clear to laypeople. The study will be interpreted as an attribution study. We know that, and so does Michael, which is why he let us know.

Michael's study will undergo the peer-review process. We have practically no choice but to engage with this laborious process too. Ultimately, it is more important to ensure that nobody can play groups of scientists against each other than to go public as quickly as possible. And we will be sticking to our own standard: no real-time attribution studies for new event types.

Nobody is entirely happy with this decision. We have a result, and while this is, strictly speaking, a new category of event, heavy rainfall associated with a tropical depression (as in Louisiana) and heavy rainfall associated with a hurricane (Harvey) are not that different.

Our reticence will cost us: for quite some time, we will be leaving the debate to those whose agenda is not driven by data and facts. And by the time we present our results, people may not be interested anymore. Nevertheless, I am convinced—and will later be proved right—that Harvey was such a disaster for the U.S. that nobody will forget it anytime soon.

In addition, the American environment is so hostile to science at the moment and so full of "alternative facts" that we need to make sure our real facts are truly watertight—even more than usual, given that hurricanes are new territory for us, and the U.S. has changed a lot since we studied the Louisiana floods in 2016.

The fact that climate change has made Harvey more likely—whether three times or four—will raise many questions that could prove uncomfortable for the powers that be in Washington, Austin, and Houston.

CONSEQUENCES

The Power of Attribution Science

CHAPTER 6

IGNORE CLIMATE CHANGE AND SUFFER ITS WRATH

ONE YEAR BEFORE Harvey submerged the city of Houston in water for days on end, claimed at least eighty-eight lives, and caused $125 billion in damage (more than any natural disaster in U.S. history except Hurricane Katrina),[1] Mike Talbott gave a remarkable interview. The longtime director of the Harris County Flood Control District, which also encompasses Houston, was interviewed by journalists from the *Texas Tribune* one more time before his impending retirement. Asked whether the authority incorporates climate change into its plans to protect Houston, Talbott responded: "I'd love to, if somebody would tell me what that is." He continued: "Give me a number. What's the number I should be using instead of the historical numbers?"[2]

He went on to state that climate change had not been discussed specifically and that, in any case, he did not consider heavy rainfall to be the "new normal"—unlike many of the scientific and environmental experts of whom Talbott alleged that "their agenda to protect the environment overrides common sense in a lot of cases." His verdict? "They're anti-growth."

Talbott cannot be accused of failing to do his job and surrendering the fourth-largest city in the U.S. to the floodwater. In his thirty-five years with Harris County Flood Control, he strove to better protect Houston by expanding canals and river branches. He did, however, reject more drastic measures such as stemming the city's unbridled growth and increasing the number of floodplains.[3]

Houston has truly lived up to its nickname of the "city with no limits": from 1995 to 2015, the population rose by a quarter to 2.2 million. Houston is the only city in the U.S. with no zoning laws—people can build more or less wherever they want. Constituents have regularly voted for this. New buildings have even been erected on floodplains—areas designed to soak up floodwater.

This policy took its toll at the end of August 2017, when Harvey dumped masses of water on Houston, and the U.S. Army Corps of Engineers decided to flood a westerly district to stop two reservoirs from overflowing, preventing an even greater disaster in the inner city. This district had actually been designed to collect floodwater, but since extreme rainfall caused by storms like Harvey was estimated to occur only around every hundred years, this possibility had been ignored and some areas had been developed.[4]

The water came at night, inundating streets and houses and leaving many residents with practically no time to prepare. Some elderly people reported waking up in confusion, thinking they were in waterbeds.[5]

Among the areas to be subjected to controlled flooding was one of the city's richest districts, home to many of the people working for oil companies like BP, Shell, and Exxon-Mobil—companies whose business models increased the volume of water carried by Harvey.

In addition to the development issues, flood experts criticized the fact that the drainage system was not designed to handle a massive storm.[6] While better planning could not have prevented Harvey from inflicting major damage, it would at least have curbed it significantly.

"Houston is the Wild West of development, so any mention of regulation creates a hostile reaction from people who see that as an infringement on property rights and a deterrent to economic growth," Sam Brody, director of the Center for Texas Beaches and Shores at Texas A&M University, explained to the *Washington Post*.[7] "The stormwater system has never been designed for anything much stronger than a heavy afternoon thunderstorm."

Houston reflects one American mindset that is hostile to regulation in any form. It has even made it to the White House: the Trump administration rejects climate protection as anti-economic and has even had the concept erased from government documents, as though one of today's greatest environmental problems could simply be dismissed.

Mike Talbott is in good company. Put simply, many city planners have failed to confront the new challenges presented by climate change.

When people draw up plans—whatever the subject and scale—they focus on risks and costs. They need to know which risks they are willing to take. To do this, however, they need to know what the risks actually are. It is understandable that planners might choose to equip their cities only for events that occur more frequently than every hundred years and categorize all less likely extreme events as natural disasters that humans are powerless to stop. But they also need to know what form these extreme events might take. At least for Houston, the historical figures were obsolete.

Our World Weather Attribution team was able to show that climate change had made Harvey's rainfall much more likely and put this into numbers. Talbott was wrong to state that climate change cannot be calculated. It can definitely be calculated.

Our study on Harvey was published on December 14, 2017.[8] The press conference took place in the stifling, windowless rooms of the New Orleans Convention Center, the venue for the annual meeting of the American Geophysical Union, the world's largest climate research conference. Karin represented the World Weather Attribution team and was joined on stage by Michael Wehner, who had also produced a study on Harvey. This was Karin's first press conference and she was suitably nervous, although you couldn't tell. She summarized our findings like a seasoned professional: Climate change had made the rainfall over Houston much stronger. In summer 2017, more than 39 inches of rain had fallen over Houston in three days—12 to 22 percent more than would have fallen in a world without climate change. Therefore, climate change had made a significant difference, even if Harvey would probably have laid waste to the city without it.

Three further studies reached similar conclusions; these had been conducted by scientists who had studied Harvey completely independently of one another using very different methods.[9] It was a scientific triumph—particularly for us, as it showed that the World Weather Attribution team wasn't just a group of lonely eccentrics, but scientists who knew what they were doing.

However, extreme rainfall like that of Harvey has become not only heavier, but also more frequent. The probability of heavy rain as seen in Houston in August 2017 has approximately tripled due to climate change—increasing more than we could have expected just from the link between a warmer atmosphere and greater rain capacity.

While Harvey remains a very rare, extreme event, it should give city planners food for thought—and not only because it will grow more likely with every further 1°C (1.8°F) of global warming and might not be all that unusual at some point in the distant future.

No, Harvey hints at a far more immediate threat. To explain this, I need to start with Emil Julius Gumbel, a publicist and radical pacifist born in Munich in 1891. As well as staunchly defending the Weimar Republic, Gumbel made a name for himself in mathematical circles through his work on statistics. The Gumbel distribution, which describes a statistical particularity, was named in his honor. As we discovered, extreme rainfall in many parts of the world—and in Houston too—follows this distribution: in contrast to seasonal averages (for example), heavy rain in Houston always behaves in the same way, regardless of how rare the event is. This means that if an event that occurs every thousand years becomes three times as likely, so does a once-in-a-century event (although greater uncertainty surrounds thousand-year events due to scarce data). Therefore, once-in-a-century rainfall of 4.13 inches per day in a world without climate change now occurs every thirty years or so. While this might not be as intense as Harvey (13.97 inches per day), it would probably still be enough to trigger a disaster.

Our study was mentioned in more than a thousand media reports, many of them highly detailed and well informed, such as those in the *Washington Post* and *New York Times*. Even *Breitbart* ran an Associated Press article that quoted Karin, who, when asked whether climate change had made Harvey more likely, answered with a clear "yes."[10] All this attention raised public and political awareness that climate change is real and is taking place here and now.

Houston's city planners and decision makers could have really used a study like ours before Harvey arrived. And yet the preceding years had certainly brought many warnings that climate change was increasing the intensity of rain in and around the city.[11] The series of tropical storms that had afflicted Houston since 1998 might also have set them thinking. A few years before Harvey, hundreds of homeowners in the city and surrounding area sought compensation for flood damage caused by three tropical storms between 1998 and 2002 alone. The court rejected their lawsuit by a majority, pointing out, among other things, that even by the homeowners' accounts, the flooding had several causes including "acts of God."[12]

Under the reign of Talbott and his successor, recent developments in climate change were simply not incorporated into planning. Calculations of events likely to occur once a century were based on data from the middle of the twentieth century, a time when the average global temperature had increased by only one or two tenths of a degree.

Waiting Until Disaster Strikes

Houston is not an isolated case. After Hurricane Katrina hit New Orleans in 2005 and set the record for hurricane damage so early in this millennium, you might have thought that U.S. cities and states would align their flood protection measures with climate forecasts. But any action that was taken wasn't initiated until after disaster had struck. It wasn't until Hurricane Sandy that New York moved its emergency hospital generators from basements to higher stories. North Carolina also waited too long.

With its long, low-lying coast, North Carolina is particularly vulnerable to storms. A few years ago, its Coastal Resources Commission developed a worst-case scenario that predicted a sea level rise of 39 inches over the next century. Politicians responded, although not as you might expect: in 2012, North Carolina's government passed a law prohibiting policies based on such forecasts.[13]

Why? Because building contractors feared that real estate and property values would drop dramatically and insurance premiums would rise. Rather than adapting to the increased sea level expected in the worst-case scenario, politicians decided to classify this as a much lower risk—based on historical data. Little changed under the new governor, Roy Cooper, despite his announcement that North Carolina would join the alliance of American states committed to the goals of the Paris Agreement. "Instead coastal development flourishes as more beachfront buildings, highways and bridges are built to ease access to our beautiful beaches," coastal geologist Orrin Pilkey wrote in the regional daily *News & Observer.* "Currently the unspoken plan is to wait until the situation is catastrophic and then respond."[14]

It didn't take long: in mid-September 2018, Hurricane Florence submerged large portions of the state (along with South Carolina and Virginia). Dozens of people lost their lives in the floods, and experts estimated the damage at $24 billion.[15]

If city planners were always aware of all possible risks, there would be no need for attribution studies. Everyone would know the critical events and would consult climate simulations and weather data in appropriate measure. However, most people, understandably, don't think about the risks or realize their own vulnerability until something happens. Climate change only becomes interesting when it affects one's own interests. And most people's interests are restricted to their own location and time. There has always been extreme weather; of course there has. But when an attribution study exists, nobody can shrug off a disaster as mere bad luck.

Europe is no exception, as the record-breaking heat wave of 2003 shows. Hospitals and authorities were not at all prepared. Given the high number of victims, there was no avoiding the question of whether this was just weather or whether the changing climate also played a role. One year later, the first attribution study provided the answer: climate change had made this sort of heat wave twice as likely.[16] Sweltering summers were here to stay.

Europe has actually adapted in the meantime. When France was hit by a similar heat wave in 2006, far fewer people died than three years before. However, the authorities' response was largely due to their shock at the high mortality rate in 2003, not the attribution study. When the study was released in 2004, it did at least provoke a discussion in Europe about climate change and heat—based on specific figures, which may have emphasized the need for better preparation. For example, the foreword to the heat wave plan drawn up by the British National Health Service in 2004 cites the figures in this attribution study.

Surprised by the Vagaries of Climate Change

However, the case of the Netherlands shows that plans must be continually updated to accommodate the evolving consequences of climate change. Since 2003, whenever summer temperatures rise, the Dutch authorities have regularly reminded people to stay hydrated, avoid direct sunlight, and, in case of doubt, to sleep on the sofa if their bedroom gets too hot. From June to August, these reminders and measures help to minimize heat victims.

But then came September 2016. For one week, day and night, cities like Amsterdam and Utrecht experienced extremely high temperatures. Suddenly the hospitals were full of dehydrated elderly people. What happened? Had they learned nothing since 2003?

While the authorities responsible for heat waves had learned some lessons, they had only planned for the meteorological summer. Once the season ended in August, they saw no further risk of heat waves; there were simply no plans in place for September. The September heat was another wake-up call, teaching us to define heat not by the calendar, but by actual temperatures. An attribution study (in Dutch) by my colleague Geert Jan of the Royal Netherlands Meteorological Institute did succeed in emphasizing how important this was for the Netherlands.

If you want to prepare your area for the capricious weather of the future, you need to know exactly how climate change will take effect. Paying little or no attention can prove fatal, but overestimating its impact can have serious consequences as well.

In summer 2018, U.S. researchers writing in the scientific journal *Nature Climate Change* described the potential repercussions of knee-jerk responses to extreme weather events.[17] They described cases in which short-term, highly localized measures can prove counterproductive in the long term: for example, dikes repaired immediately after floods, even though their height needed to be completely altered or the water management system simply didn't work. Or cities and districts investing heavily in sandbags even though flooding was extremely rare for them and their regions were much more susceptible to totally different weather events, such as droughts—which were, in turn, ignored.

Attribution studies do not universally protect against knee-jerk reactions and maladaptation, as it is known in specialist circles—adaptation measures that are either superfluous or that heighten, rather than reduce, vulnerability in the long term. However, attribution studies—particularly those made available shortly after an event—can help planners better assess the actual risk by calculating the likelihood of an event in the current climate and then gauging whether and to what extent this has changed and will change in the near future. If a dike is unable to withstand an event that occurs once in a thousand years, totally different measures are required than if a more frequent event would cause it to collapse. A different response again is required if a weather event becomes less, rather than more, likely.

There is plenty of need for attribution studies, no doubt about it. To date, however, governments have rarely used these studies to adapt to the consequences of climate change. It may be that our discipline is still too new, that it is still fairly unusual for scientists to deliver information quickly. We will know that our work is truly successful when it is established in the minds of the public and decision makers alike, and we are regarded as boring instead of wacky. This will probably happen faster than anticipated.

The Plan: A European Attribution Team

In October 2017, representatives of European weather services and scientists from our field met in the basement of the Czech Ministry of Transport in Prague's Old Town. Their goal was to develop a permanent European attribution service that would examine every possible case of extreme weather to determine the role of climate change.

The meeting was organized by the Copernicus Climate Change Service, which develops climate forecasts for the European Commission and national governments and proposes ways in which European nations can adapt. The meeting in Prague was supposed to clarify whether a European attribution service was actually necessary. Instead, the discussion focused on when and how the idea could be implemented.

Ideally, the service would start work in the near future, focusing on nothing but the causes of weather events and delivering results within days—even on public holidays. In other words, it would be constantly searching for fresh and immediate answers to the impact of climate change on Europe's weather.

Initially, studies would inevitably be restricted to heat waves, cold spells, and large-scale extreme rainfall. Before we can tackle wind, droughts, and storms, we need to gain more experience with the methods, incorporate better models, and, in some cases, develop whole new methods. However, it is already apparent that at some point, attribution studies will become routine, a natural part of government and authority planning.

National weather services have already begun using our studies. Since we showed that it works, journalists have started calling these services after every extreme weather event to ask about the role of climate change. As a rule, those consulted need to exercise caution and stick to talking about global developments. We can, however, examine individual droughts or floods for the hallmarks of climate change, although our small team can only cover relatively few events. The resources of the weather services and our expertise would complement each other well.

In January 2018, I visited the German Meteorological Service (DWD) in Offenbach. We discussed how the DWD could begin performing attribution studies for Germany. The enormous scale of the DWD was a bit of a culture shock for me, as someone used to making many decisions herself and working in teams of two to five people. Nevertheless, we managed to find some common ground, and the DWD is now planning its own small team to attribute individual weather events to climate change within a few days. From 2020, the DWD aims to publish its findings on floods, heat waves, and rainfall immediately on social media followed by studies one or two weeks later. Paul Becker, vice president of the DWD, told *Nature* that "it's part of our mission to illuminate the links between climate and weather."[18] He went on to say that "there is demand for that information, there is science to provide it, and we are happy to spread it."

If extreme weather were to be continually examined for the hallmarks of climate change using standardized methods, decision makers would be better able to plan than if confronted by loads of studies in which a "heat wave" means something different every time; after all, we have spent most of the last five years trying to determine which methods are most suitable. In the last couple of years or so, scientists have reached something of a consensus on the structure an attribution study should take to offer the most robust results possible and to clarify specific questions—for example, regarding heat stress or temperatures. The time has come to launch operational attribution services. This would be a huge step forward for our discipline and a great opportunity to raise awareness of climate change and extreme weather among the general public.

While important for industrialized countries, regular attribution would be crucial for developing and emerging countries where extreme weather has far more dramatic consequences. Not only are these countries much more vulnerable, but they are seeing the impacts of climate change rise much faster than elsewhere.[19]

Then again, many governments in the global South often instinctively point to climate change and the historical guilt of the West whenever a storm or heat wave has disastrous impacts on their country, even though many causes can often be laid at their own feet.

What we need is clarity, and this is what our studies provide. Once we know which drivers are truly responsible for disasters in the wake of extreme weather, we can act.

CHAPTER 7

FACTS NOT FATALISM

Identifying the Causes
of Disasters in Order to Act

WHEN I FLEW into Cape Town in February 2018, the first thing I noticed was the "Day Zero" warning posters promoting water conservation. In the public restroom, I turned on the faucet but not a single drop of water came out. Antibacterial gel was provided instead. The hotel bathroom contained a bucket for guests to collect their shower water so it could be used for cleaning. And the washing machine's drainage pipe was connected to the toilet, explaining why foam came out every time I flushed.

Cape Town's water reserves were running low and the authorities had announced that the coming months would bring a "Day Zero," as of which there would be no running water at all. For three years, the region had been suffering its worst drought in over a century.

This was the reason for my visit. Together with colleagues from the University of Cape Town, our team wanted to plan a series of studies investigating the influence of climate change on South Africa and the continent as a whole. At the top of the list was an attribution study to investigate the role of climate change in the ongoing drought.

In the area around the city, irrigation had been switched off and fields were drying out, resulting in serious crop failures and millions of dollars of damage. Now Cape Town itself was also in danger of losing its water supply.

About a month later, however, at the start of April, the authorities gave the provisional all clear and pushed Day Zero back to the following year. Thanks to drastic water rationing, Cape Town had narrowly avoided disaster.

The rain returned in June and started refilling the city's six large reservoirs. What remained was unease. Was climate change to blame for the drought? If so, how should a city rightly proud of its efficient water management prepare for a future lack of rain?

At the end of July, we were able to answer this question. By running computer simulations and analyzing data from eighteen weather stations, we calculated that climate change had tripled the likelihood of a drought like this in Cape Town.

So while an event like this would be expected every three hundred years in a world without climate change, it will now occur every hundred years. This might not sound particularly alarming, but a look to the future puts it into perspective: if the global temperature increases by another 1°C (1.8°F), the probability of such a drought could increase by the same factor again. This extreme aridity would then occur every thirty-three years or so and become even more likely with every further degree of global warming—turning a once rare event into a fairly common occurrence.

Colleagues in Cape Town can now use our study to help authorities plan for the future and for water shortages. The city currently draws almost all of its fresh water from its reservoirs. It would probably be better to diversify and obtain some of its supply from groundwater, for example. In an emergency, desalination plants would also be an option.

The effects of climate change are felt most keenly in developing and emerging countries, first because of their particular vulnerability: in the poorer areas of the global South, basic house structures would barely withstand a heavy storm. In addition, authorities' drought warnings and heat wave recommendations often do not reach the population at all, resulting in far greater damage than necessary.

Second, the risk of extreme events is rising disproportionately in these parts of the world. An increase in droughts or floods can set economies back by years. For example, 2004's Hurricane Ivan destroyed all nutmeg production on the Caribbean island of Grenada, devastating one of its most important exports.[1]

Information about the actual change in risk is crucial in minimizing this risk and making a country as resilient as possible, at least if the information is made available when needed—straight after the event, when decisions are being made about rebuilding, resettlement, and compensation.

For this to happen, the information must get to the people who need it. In developing countries in particular, decision makers and journalists have very little access to attribution studies; both the weather data and scientists in our field tend to concentrate on industrialized nations.

However, as in the global North, if politicians and planners are not aware of the various climate signals and do not consider data-based evidence, they run the risk of making disastrous decisions—for example, rebuilding in areas highly likely to be washed away again the next time it rains.

Climate Change Isn't Always to Blame

Conversely, there is also a danger that the role of climate change may be exaggerated. Some political leaders (particularly, but not exclusively, in the global South) instinctively blame all damage caused by extreme weather on climate change and, therefore, on the West.* This is understandable; developing countries suffer more extreme weather than industrialized countries, and, of course, have contributed the least to climate change. The majority of greenhouse gases have been emitted by the West, which for centuries has profited hugely from burning fossil fuels. The World Bank[2] and the United Nations[3] make the same point. However, this mindset, while justified, can also prove a hindrance and limit the scope of action—particularly if climate change is not a major driver of a drought or storm and the ensuing disaster is simply a case of bad luck or (to some extent) poor planning or preparation.

East Africa is a good example. In a world of global warming, climate researchers expect droughts to become more intense and more frequent, which is why East Africa's many dry spells over the past years have been commonly accepted as a sign of climate change.[4] In 2015, Ethiopia experienced one of its worst droughts for decades. Hundreds of

---

* Here, "the West" is synonymous with the traditional industrialized countries.

thousands of farmers in the north and center of the country lost their crops and most of their herds; eight million people relied on food distributed by international aid organizations.

When we talked with them, government officials and NGO (nongovernmental organization) representatives were convinced that the aridity had been caused by climate change, or at least that this was a major factor. The country had taken corresponding measures to adapt to future droughts. Following the recommendations of some studies, irrigation systems had been built based largely on forecasts of average changes (the average rainfall to be expected).[5] Satellite images also showed that farmers were watering their fields and farmland more.[6]

But was climate change really to blame? Had it significantly increased the likelihood of droughts in East Africa?

Using statistical analyses of weather data and climate models, we simulated countless potential weather scenarios and showed that the drought in Ethiopia was an exceptional extreme event only to be expected every few hundred years on average. Our attribution studies did not show that climate change was the main factor or had made the lack of rain far more likely.[7] While temperatures in East Africa are rising due to global warming (which normally increases water evaporation), evaporation rates were already very high in its dry and hot regions. If there is no water, it can't evaporate, so the link is not as clear as in other parts of the world.

But if climate change wasn't the deciding factor in the intense drought, then what was? The natural variability of the climate? Or perhaps other factors outside the climate system? Should politicians and planners in Ethiopia and its neighboring countries be asking whether they could have been better prepared? It is possible for a particular country to become more vulnerable even though drought frequency and intensity haven't changed that much.

In addition to the atmosphere, ground humidity and the status of reservoirs are also important, as are vegetation density and species. Humans can have a major influence on all of these factors, for example by clearing forests, altering their agricultural practices, or converting grasslands into fields. Of course, the level of preparation also determines how problematic a drought will be. To what extent does the population depend on the rain? When do they become aware of an impending dry spell? What reserves are available to survive a bad harvest, and to whom? And how many people are insured?

Most of East Africa's population depends on agriculture. While the people in the north tend to move from place to place depending on the season, central East Africa grows lots of corn—a field crop that only thrives if it gets enough rain. Farmers who focus mainly on corn are therefore particularly susceptible to drought. Droughts have always been a fact of life for the people of East Africa and yet they have continued to grow corn, despite the risk of veering between no harvest and abundant crops year by year.

On one of my visits, Dr. Abiy Zegeye of Addis Ababa University explained that a corn farmer enjoys a lot more social prestige than someone who grows millet or other, less exciting, basic foodstuffs.

Rice is another important field crop in southern East Africa. While it copes relatively well with high temperatures, it also needs a comparatively large amount of water—making it vulnerable to droughts.

Over the past decades, Ethiopia has also cleared a large proportion of its forests. The government is now turning its attention to remote regions, distributing land rights and building roads.

To counteract the droughts, government authorities began to build irrigation dams. However, these triggered floods that destroyed pastures and forced the nomadic Afar herders to move to other areas, in turn fueling conflicts and reinforcing social tensions.[8] While this was not the government's intention, and the irrigation system may have provided more water for the local population, its actions actually increased vulnerability in times of drought.

Politicians have a tendency to blame climate change and to tackle the drought, rather than working to make a region more resilient. They fight the effect, not the cause. Weather is seen to play an overwhelming role, and well-intentioned reports from the World Bank (for example) lead them to place the blame on climate change. They may truly believe that climate change is the region's biggest problem, but it can be convenient to have someone or something to blame.

Our studies on Kenya, Ethiopia, Somalia, and East Africa as a whole have shown that climate change was not the main reason for the lack of rainfall.[9] This doesn't mean it wasn't involved somehow; a lack of rain is not the only thing that makes a drought. If the information we obtained about the Kenyan drought of 2017[10] had been available to decision makers in 2010 for Ethiopia, and attribution researchers worked more closely with social scientists, different measures might have been available in the Afar region: measures that focused not just on a specific meteorological event, but also on the social and political structures that largely caused the high economic and social losses.

Improving education can sometimes be more useful than building reservoirs. Developing countries are not the only ones to focus on the wrong things; foreign development organizations will often arrange for a well or reservoir to be quickly constructed to use funds that will dry up at the end of the financial year. Long-term projects are difficult to finance. Attribution won't eliminate this long-standing problem, but if we can show that a certain weather pattern is no coincidence—with specific data tied to real experiences—then we can at least improve awareness and, hopefully, help funding and development projects to be deployed more efficiently.

Politicians on the Spot: Inaction Is No Longer an Option

What happens if we show that climate change was not the main culprit for once? On a positive note, this gives power to decision makers; their actions to develop a region will not be immediately ruined by more intense and frequent droughts, for example. It also imposes an obligation on decision makers. Failure to act despite the new evidence would be dereliction of duty, if, for example, a government does nothing to alleviate the consequences of droughts or other natural events by making its people more resilient.

This might sound drastic, but it is crucial to emphasize that the failure to build resilience is not a deliberate act.

This research is not about apportioning blame or absolving the West, but about determining facts, exploring potential action, and giving agency to local people. I assume that even autocratic governments have a vested interest in making their people less vulnerable. But even if the will is there, the opportunities may not be. The political structures required for action are often lacking (not exclusively in Africa, of course), along with the data that will enable people to prepare themselves for extreme weather.

The Kenyan government is one example of the need for these data; the politicians were very interested in our attribution study on the 2017 drought, even though the result, once confirmed, was not easy to communicate.

The dry period not only affected parts of Kenya, but caused havoc in Somalia and sent a wave of refugees fleeing to Kenya. When it became clear that climate change was not (as expected) a major cause, however, enthusiasm about our findings was muted as government employees and NGO representatives realized that communicating this was going to be tough.

Preparing for the press conference was suitably tricky. We spent many hours sitting in the garden of our hotel in Nairobi, honing every last word. At the very least, it allowed us to escape the dust and noise of the city and enjoy the shade of palm trees full of frolicking monkeys.

While I would have preferred to emphasize that climate change was not the main reason for the lack of rain, the headline of our press release stated that we were uncertain as to the temperature's role in the region's droughts.

Along with government representatives, we presented the results in Nairobi in March 2017 as the drought continued in Somalia. Despite the initial difficulties in presenting the results, it was a great success for our discipline and science as a whole. The government's participation shows that attribution studies are important and welcome, even when the results do not necessarily serve all political interests. It also indicated that, sometimes, governments do actually want to understand the true causes of weather events to ensure that they take the right actions.[11]

Attribution studies can encourage development and climate change to be considered together, rather than being played off against each other—particularly if a study is performed at the right time (i.e., during a drought or immediately after a flood). Studies produced wholly or partly by local scientists, like the study on the drought in Kenya and Somalia, are essential, though. Nobody understands the local meteorology and environment like the locals, and there is nothing worse than a few European scientists turning up and claiming to know everything about the place.

We work closely with the Kenyan weather service—most recently in spring 2018, when Nairobi was battling floods. In South Africa, too, we are working with scientists from the University of Cape Town on a new project that builds on our experiences in Kenya to explore how Africa can effectively adapt to the extreme weather of the future.[12]

All around the world, our studies have shown that in some cases climate change plays only a minor role in extreme weather events compared with other driving factors. Even if an event bears the hallmarks of human-caused climate change, other factors will also be important. Automatically naming climate change as the sole cause of disasters is, therefore, implausible and fails to recognize the reality of a complex system with many intertwining social and physical factors. Conversely, to reject climate change entirely is to misjudge reality.

In an increasingly complex world, simple answers can be seductive. Environmental and development organizations often simplify connections to gain easier access to donations. This is understandable to an extent and is part of their work, but sometimes headlines can be a bit too misleading. For example, Oxfam published its own press release shortly after our press conference in Nairobi. The results of our studies were included in the document, but the headline read: "A climate in crisis—how climate change is making drought and humanitarian disaster worse in East Africa."[13]

The organization undoubtedly wanted to help the people in the region, and there is no doubt that temperatures are rising and that this can be attributed to climate change. However, the connection between rising temperatures and these droughts is speculative at best.

Fleeing the Climate and Weather

For some years, people around the world have been discussing the influence of climate change in Africa and Southeast Asia in another context too: "climate refugees" are now often mentioned by the media, which already focus heavily on refugees.

However, there are currently very few indications that increasing numbers of people are actually migrating (yet!) due to weather or climate events. When people are forced to leave their homes due to droughts, floods, and other natural disasters, it is not known whether they do this because climate change may have amplified these events; at least, there are no attribution studies on weather events for which refugee figures are available.

A person who decides to leave their home in Africa will usually do so because of military conflict or other political or economic factors; even weather refugees, some of whom are climate refugees, will often leave the region but not the country itself, and in many cases will return home. Climate refugees do exist, but they are more likely to move from, for example, Bangladesh's smaller towns and cities to its capital, Dhaka, or from Somalia to Kenya.

Migration is a highly complex topic and a tricky subject for scientific research. There are many reasons why a person might become a refugee. This makes it easier to see what you want or expect to see depending on your worldview, while facts that demand a different interpretation tend to be ignored. Most of the time, this isn't intentional.[14] While there may be a wide variety of methods and studies on the causes of migration and migration flows, and just as many studies on climatic changes all around the world, there is little overlap between these two very different scientific fields. Scientific evidence of when and how many people actually flee their homes due to climate change is therefore extremely thin on the ground.*

---

* Studies fall back on global, long-term estimates of the costs and impact of climate change, from sources like the IPCC reports or Nicholas Stern's *The Economics of Climate Change*.

From a global perspective, it is clear that climate change will fundamentally alter the lives of millions of people. Rising sea levels will render island nations uninhabitable with flooding during every storm. In a low-lying, heavily populated country like Bangladesh, many people may be forced to modify their livelihoods and leave their homes, as the next chapter will show. Climate refugees will exist; they may already exist, but there are no reliable figures to show whether they are permanently leaving their home countries or continents and how many of them there will be.

Together with social scientists, we can use our attribution studies to try to shine a little bit more light and bring facts to a debate that largely ignores the reality of refugee movements. Lisa Thalheimer, one of my students, plans to do just that. She has already worked for the World Bank on East Africa. For the first part of her doctoral thesis, she has analyzed all studies of weather and migration in East Africa to see whether they make a connection with climate change. The initial conclusion is that they do not. In most cases, the link between migration and a weather event (or several events) has not been quantified or documented.

Next, she is looking at weather data and migration databases in order to link them with attribution methods. It is not yet clear whether these new, systematically collected data can be used to establish a causal link between migration and climate change.

From a scientific perspective, it is not surprising that there is no simple, causal link between climate change and migration. Scientists at the University of Hamburg have studied potential climate refugees in various Asian countries and West Africa on behalf of Greenpeace. Their study was published in May 2017.[15] They too found that when extreme weather triggers a natural disaster, people only flee if other political and social factors come into play. The study also makes it clear that the number of people making their way to Europe is extremely low compared with those who sit tight or move to another region or a neighboring country until the situation improves.

However, the study was only able to discuss the connection between these weather events and climate change in general terms—for example, stating that more heavy rain and floods are to be expected as a global average. No specific numbers were provided for the cases featured. We now have some of these numbers, for example on the two events captured in photos in the Greenpeace report: a flood in Thailand in 2011 and a drought in the Indian state of Maharashtra, which led to significant crop failures in 2016. In the case of Thailand, we were able to show that climate change had not changed the probability of extreme rainfall, the cause of the floods.[16] We do not yet have a definitive result for Maharashtra and are currently working on an attribution study with the Indian Institute of Technology Bombay. Preliminary results suggest that climate change was a contributing factor in this drought, which in turn sent a wave of people fleeing to Mumbai.[17]

In many cases—such as the Thailand example—these supposed climate refugees do not exist. In other cases, such as Maharashtra, a lack of social infrastructure, as well as political instability and inefficiency, means that climate change amplifies disasters resulting from weather events.

In other words, climate change must be stated as the reason why people fled, even though the situation was determined by totally different problems that have been overshadowed by the "climate refugee" argument. If we can shed light on this issue, then one battle has been won—particularly in a debate largely free from facts and fueled by emotions.

This chapter has been the hardest to write. I am aware that everything I have said over the last few pages could be terribly misconstrued. And yet the content of this chapter is one of the main reasons why it is important to attribute individual extreme events like droughts and floods. If we can pinpoint when climate change did or did not play a crucial role, then we can help to solve problems.

In no way am I suggesting that industrialized countries should stop worrying about the problems in the global South simply because climate change, for which they are largely responsible, is not the main culprit in every single case. As the historically greatest source of emissions, industrialized countries are accountable and must relieve developing countries of some of the burden they carry today and will carry in the future. We are now able to determine, though, where climate change really is a game changer, and where other causes dominate. As the next chapter will show, this makes a huge difference.

CHAPTER 8

A QUESTION OF JUSTICE

The Cost of Climate Change and the Responsibilities of Industrialized Countries

ANGLADESHI SCIENTIST Dr. Saleemul Huq doesn't necessarily look like someone who's about to change the world. Reserved but friendly, he seems unassuming at first, or at least he did when we shared a podium at the 2016 global climate conference in Marrakesh to discuss his life's work: Who should pay for the increasing climate-related *loss and damage* (as it is known in conference jargon) seen throughout the world?

This loss and damage occurs because the world has relied on fossil fuels for too long, allowing human-induced climate change to develop consequences that cannot be or are not being prevented through adaptation: torrential rain washing away entire houses, storms pushing seawater onto the land and ruining crops in coastal fields, and people dying in extreme heat.*

Dr. Huq is an old hand at global climate conferences and has participated in all twenty-five summits so far. Among other things, he encourages the world's poorest countries to go on the offensive, advising them that such damage has long since become normal and that the people behind it— the industrialized countries—need to pay.

Most people don't dare discuss the topic openly, but he doesn't mince words, which is probably why scientists, advisers, and politicians all flocked to him after our podium discussion, eager to exchange a brief word with the influential director of the International Centre for Climate Change and Development in Dhaka.† This happened every day of the two-week conference, and my colleagues reported the same at previous conferences in Paris, Lima, and Warsaw.

Dr. Huq knows what he's talking about; after all, his country is facing a huge problem. Not only is Bangladesh one of the world's poorest countries, it also has one of the densest populations, with around 165 million inhabitants over 57,915 square miles—half the size of Germany with around double the people. One in three people live by the coast, where the Ganges and its tributaries form a delta with sprawling, fertile fields and green forests. This region

---

* There is no consensus on what constitutes "loss and damage." Most scientific articles on the topic talk about loss and damage that has not been prevented by swift reductions in greenhouse gases by the global community or by measures to adapt to climate change.

† Dr. Huq is also a senior scientist at the International Institute for Environment and Development in London.

is particularly vulnerable because it is flat and low, a suitably large target for the ocean, which is slowly swelling thanks to global warming.

Bangladesh is used to water; it depends on it. A quarter of its total land surface is submerged at least once a year when the monsoon rains descend and rivers leave their beds. This flooding benefits its many rice fields, which are nourished by the soil the river brings. Rivers are so important that they are typically described as the "mother" of the country.

But if the mother starts to drown her children, then there's a problem. Since the 1960s, Bangladesh has been taking measures against this, reinforcing riverbanks and strips of coastline with all manner of dams and barricades, particularly in the southwest. Sometimes, however, these exacerbate the problem by preventing the traditional inter-action of water and land: if the river water cannot distribute its sediment over the land because the canal and river walls are holding it in and building it up, then the land to its left and right will sink. Even at normal water levels, the surface of the water will be higher than the land, like a full bathtub. If a cyclone destroys the dams, severe flooding may wash away huts, people, and animals with waves of mud. In 2009, for example, Cyclone Aila pummeled the country with storm surges, breaching several embankments. More than 150 people died and the damage amounted to $270 million. Geologists visiting the affected area determined that the marshland outside the dams was over a foot lower than the average floodwater level.[1]

Still, the country responded as usual: by building even higher dams, with $400 million of funding committed by the World Bank. However, some farmers have learned from the mistakes of the past and have started to experiment with the opposite approach—controlling water inflow through targeted gaps to relieve the pressure and stop the land from sinking.

Bangladesh is exposed to potential natural disasters on all sides: fluctuating monsoon circulation, floodwaters breaching riverbanks, storm surges caused by tropical cyclones. In such a poor and vulnerable country, even a slight increase in the intensity and frequency of extreme weather will have serious consequences. One of our most complex attribution studies to date focused on the flooding of the Brahmaputra delta in 2017. We concluded that climate change had made the rainfall that caused the flooding around 70 percent more likely.[2]

Current events are just the beginning. The rising sea level will pose a real threat to the country's survival: the sea level off the coast of Bangladesh is forecast to rise up to 4.9 feet by the end of the century.[3] Calculations show that this would eat up 16 percent of its land surface and force millions of people to relocate.[4]

Rich Countries Don't Want to Change the System

Wherever you look, it's obvious that climate change has led to loss and damage: Bangladesh, the American east coast, the small island states, and Indonesia, which, with vast parcels of low-lying land, was hit particularly hard by the tsunami of October 2018. So it's all the more astonishing that this problem was barely discussed for such a long time. For years, the topic played no role at all at climate summits. As so often, this first comes down to money.

Loss and damage means high costs, costs that someone has to pay. And since these costs are largely incurred by countries that have contributed little to climate change, this raises the question of justice and responsibility—a question that, for this very reason, the main contributors to climate change do not want to answer under any circumstances. People who profit from a system often don't want to change it; rich states and fossil-fuel-exporting countries benefit from fossil fuels. They say the right things and then prevent necessary measures from being taken. This is one of the main problems with climate negotiations.

I am reminded of these main contributors every morning when I get to work and make a coffee. A poster on the kitchen wall shows two footprints comprising circles of varying sizes.[5] Countries with particularly heavy carbon dioxide emissions are given suitably large circles, while countries with light emissions have small circles. In the footprint that represents total carbon emissions by nation, the U.S. forms the heel, China most of the ball of the foot, Europe and Russia the arch, and India the big toe. All other countries have smaller circles; the African nations are barely visible.

Admittedly, this graphic is almost ten years old now, and China has overtaken the U.S. as the greatest source of emissions (if looking at annual emissions). Nevertheless, the image clearly shows who has been growing rich by burning fossil fuels. The larger the circle, the greater the responsibility, especially considering the historical dimension.

The second reason why climate-change-related loss and damage was long considered a problem of the future is connected to attribution science: it is only in the last few years that we have been able to attribute specific extreme events to climate change—and thus to determine and quantify the actual damage caused by climate change. As of 2018, the scientific journal *Nature* has counted around 190 cases in which climate change has been observed to influence events that have caused real damage in the here and now.[6]

Is it a coincidence that international politics began to take the subject seriously and take action in the last few years as well?

For many years, the small island states in particular have attempted to build pressure. For many of these countries, climate change is no uncertain, incalculable issue to be tackled in the distant future, but a question of survival. To advance their cause, they formed the Alliance of Small Island States and added loss and damage to the agendas of global climate summits. They have also called for global warming to be restricted to 1.5°c (2.7°F) and to be compensated for their loss and damage by the people who have caused climate change. For a long time, they were ignored.

This changed in 2013, when climate diplomats decided at the (otherwise unremarkable) climate summit in Warsaw that, in the future, they would in some way attend to the damage caused by rising sea levels and extreme weather events.

The small island states, and people like Dr. Huq, achieved their actual breakthrough in December 2015 at the climate summit in France. The world's nations had reached an agreement on the Paris climate accord, which they all ratified or accepted. The global community committed itself to limiting global warming to 2°C, or even 1.5°C if possible; for some island states, the margin between these two values is the difference between survival and destruction.

The agreement also explicitly acknowledges loss and damage as a consequence of climate change, the importance of recognizing this loss and damage, and preventing it where possible.[7] This passage can also, if one is very generous, be seen as recognition of our work and an indirect mandate for attribution science.

We would not have expected such success; for quite some time, the industrialized countries refused to engage with the topic at all. Ultimately, it simply could not be ignored—the loss and damage caused by climate change has become too obvious, and while loss and damage means something different for almost everyone, we are now able to calculate it for individual cases using one very specific definition.

A Global Problem Nobody Wants to Name

Industrialized countries aren't putting their heads too far above the parapet just yet. While they acknowledge climate damage, they have added a special clause to the Paris Agreement expressly ruling out any form of compensation for loss and damage caused by climate change.* So while the agreement specifically determines that loss and damage will not be compensated, the actual definition of *loss and*

---

* Paragraph 52 of the declaration to the Paris Agreement expressly states that the support for loss and damage mentioned in Article 8 does not include compensation.

damage remains pretty vague. This is where things start to turn Kafkaesque.

An older UN document talks of the "actual and/or potential manifestation of impacts associated with climate change in developing countries that negatively affect human and natural systems."[8] When we used this definition in a scientific article,[9] we were told by people working on loss and damage within the UNFCCC (UN Framework Convention on Climate Change) that this is in no way an official definition, that no official definition exists, and that no meetings or official negotiations are being held to develop a definition. We were asked to publish a clarification to this effect.

We fulfilled the request, if with some astonishment.* From a scientific perspective, I see little sense in conducting

---

* Informed that no definition exists, we attempted—led by social scientist Emily Boyd—to establish what the various parties in the UN climate negotiations believe *loss and damage* to mean. We found four interpretations. Of these, the definition that most closely resembles the idea originally conceived by the small island states is "irretrievable loss of habitat, culture, and other essentially immaterial assets destroyed due to man-made climate change." This is opposed by an interpretation that cannot really be separated from the consequences of climate change to be addressed by adaptation measures: "changes in the risk level of certain weather events or environmental risks, independent of climate change or in any case definitively independent of scientific evidence or a clear causal link between these changes and anthropogenic climate change." According to this interpretation, it is difficult to separate money for loss and damage from money for adaptations to climate change or general development measures. Both interpretations and the nuances in between are almost consistent with the Paris Agreement, except for the explicit rejection of compensation. Therefore, the interpretation that perhaps most closely resembles the original idea is least compatible with the wording of the agreement. (Boyd et al., "A Typology of Loss and Damage Perspectives.")

official negotiations on a topic when nobody knows exactly what it means. If I had become a diplomat or politician rather than studying philosophy and physics, I probably wouldn't have been so surprised. It makes plenty of political sense to keep a phrase like *loss and damage* as vague as possible, particularly when the people at the negotiating table have widely varying interests. If a clear definition had been established, the concerns of the island states and other developing countries probably wouldn't have made it into a document as important as the Paris Agreement in the first place.

The more we focused on this topic, the clearer it became that we had opened a real can of worms. When I discuss my work with people in developing countries, loss and damage is often the first thing that comes to their minds—whether I'm drinking tea with ministers in Delhi, meeting science journalists in Nairobi for a working breakfast, or talking with students in Addis Ababa. In most cases, they immediately ask me not to make it an official topic of discussion. It's simply too controversial.

Everyone wants to let sleeping dogs lie. As long as *loss and damage* is not clearly defined, the people causing climate change can sign agreements that will at least move the issue a little further forward. Developing countries won't see any money from it immediately, but the mere fact that loss and damage has become a separate pillar in the architecture of climate negotiations, and is thus being acknowledged, is a provisional success and a foundation upon which to build. This may be the first step toward greater climate justice.

The Power of Numbers

By developing methods that allow us to identify climate-change-related damage done by extreme weather, we may be supplying the missing piece of the puzzle that will make the people responsible for climate change fulfill their duties and assist those who are suffering.

It's unlikely that anything will change overnight at a global level, but the carbon price concept shows the power that numbers can have, even if it takes a while.

The idea of viewing carbon as a part of production costs, and thus as a way to price the effects of climate change in line with market economy laws, has been around for many years. Economist William Nordhaus developed the concept of the carbon price in 1975 and received the 2018 Nobel Memorial Prize in Economic Sciences for his work.[10] However, companies have only started to explore ways of reducing their emissions since economists learned to realistically calculate how much a ton of carbon emissions needs to cost to cover the resulting environmental damage. Businesses around the world are now taking the carbon price seriously, even if they rarely need to pay at present and there is still a long way to go before "taking it seriously" evolves into effective pricing.

More and more jurisdictions are introducing such a price—Europe, Canada, and California have already done it, and China and Mexico are on the brink of implementation. The fact that the carbon price can now be specifically defined is often enough of an incentive for companies to adapt to future costs. The carbon price is not yet high enough to make burning fossil fuels an unprofitable business model, but more than 1,400 companies are now pursuing policies that treat this as a real possibility, even in countries where there is no carbon price.[11] This is not yet enough to effectively fight climate change, but it shows the impact concrete numbers can have.

Attribution research could help the study of climate damage take an equally large step forward. If climate damage can be translated into economic damages[12] and everyone knows the amount of money at stake, politicians worldwide will be pressured into finding a solution.

Journalists from the Energy and Climate Intelligence Unit in the U.K. have done precisely that, scrutinizing fifty-nine attribution studies from 2016 and 2017 and selecting the forty-one cases in which climate change made extreme weather more likely. For their *Heavy Weather* report, they then calculated the total damages incurred through climate change.[13] They worked on the simplified assumption that if climate change has doubled the likelihood of an event, then half of the costs resulting from the event should be attributed to climate change. This is, of course, a major simplification—damages do not increase in line with the frequency of the event—but using rough calculations to gain an initial impression of the scale of climate damage is a legitimate approach.* Using this method, they calculated that climate change was responsible for $1.6 billion of damage in the South China floods of 2015.

---

* The attribution studies on which these numbers are based were never intended as the basis for such calculations. If this goal were defined from the outset, the calculation methods could certainly be improved.

The authors didn't just look at the financial costs of climate damage; they also examined the numbers of victims. For example, the 2015 heat wave in India and Pakistan killed almost 4,000 people, at least 2,800 of whom could be attributed to climate change—an unbelievably strong argument for action. If these facts are made known before the next heat wave, then a country could be held liable if it does not do enough to adapt to climate change and its harbingers of doom, extreme weather events—if it forges ahead in full knowledge of the resulting loss and damage.

Concentrating on Risks, Not Damage

It will probably take a few more years before the world's nations agree on a mechanism that actually deals with the damage taking place in the present, and so some countries are taking matters into their own hands. For example, the Bangladeshi government has launched a national mechanism, if not to compensate for loss and damage caused by climate change, then at least to help eliminate it. It has already put money aside for this purpose. Dr. Saleemul Huq, who also advises the government and writes a weekly column in the *Daily Star*, Bangladesh's biggest daily newspaper, has certainly played his part.

However, the mechanism has not yet entered into force. Some aspects remain unclear: Under what circumstances will money be released from the fund, which is financed through taxes? And for which types of damage?

This question in particular reopens the dilemma that haunts climate summit negotiations. My position is that it would be logical to only include damage that can be attributed to climate change. Otherwise, where will we draw the line? Should we also include damage caused by a cold spell that climate change has made less likely? Or only damage from extreme weather that is happening more frequently? This would exclude some droughts, such as the one in São Paulo. The drought caused serious damage, and climate change did play a role. The area around the city dried out faster due to the heat, but it also rained more. The two effects canceled each other out, and so climate change did not make the drought more frequent.

And what about disasters caused by humans, such as a bridge collapse after heavy rain, that have more to do with lack of maintenance?*

Damage can only be specifically linked to climate change with the aid of attribution studies. Without attribution, we cannot distinguish between climate damage and damage caused by other factors.

Attribution works well for many events, but not for all: events that we cannot yet attribute to climate change include small-scale flash floods due to heavy rain, hailstorms, and tornadoes. Floods caused by short and localized heavy rain are crucial for Bangladesh. We have a lot of work to do if we want to achieve a more or less complete inventory of extreme weather events.

Furthermore, who should conduct these studies? The

---

* Damage due to natural disasters, regardless of their cause, tends to fall under the purview of other agreements, such as the Sendai Framework for Disaster Risk Reduction.

government that requests assistance? The countries set to pay for it? In either case, suspicions could be raised that a study has been steered in a particular direction. The task could, therefore, be passed to a neutral organization like a national weather service. But even that would be problematic, discriminating against poor regions or sections of the population without the necessary scientific infrastructure. Good weather records are required to achieve relatively robust results, but these are not available for every location and often not for the regions home to the poorest people. We see this all around the world: most weather stations are located near airports, military bases, or research institutes, not in slums or rural areas.

Then there's another problem: the extent of the damage after an extreme event also depends on the vulnerability of the disaster location. Places in which no precautions have been taken will, of course, suffer the most damage. It may therefore be the case that while climate change played a relatively large role in a weather event, good preparation has prevented it from causing too much damage; meanwhile, another area may experience serious damage even though climate change played only a small part in its weather. As a result, lots of compensation would be paid for sloppy adaptation—a false incentive structure.

It could make more sense to concentrate on the risk and not on the actual damage. Climate risk insurance—which has become increasingly popular in recent years—could prove a fairer mechanism.

Insuring Against Climate Risks

It works like this: if you pay a premium, you will be quickly compensated (with money or practical assistance such as seeds) if droughts, hurricanes, or heavy rainfall affects your country. A country can insure itself or its inhabitants directly. Insurance companies begin by assessing a country's risk of a specific extreme event based on climate and weather data; for example, satellites can be used to measure rainfall. This information is used to calculate an index. If the severity of a weather event exceeds the threshold agreed in the insurance contract, money is paid out automatically.

More than one hundred million people in developing countries are now insured against climate risks. To give one example, the Pacific Catastrophe Risk Assessment and Financing Initiative will pay out twelve times the premium in the event of a disaster so that bridges and airports (among other things) can be rebuilt after torrential rain or tropical storms.

The insurance companies employ parametric procedures, which allow them to make payments straight after a drought or flood. Rather than waiting until the total damage has been determined, which can take weeks, they can pay out when droughts occur that exceed a specific extreme index—for example, a drought to be expected every twenty years or more. In this type of insurance, it is significant if an event that previously occurred every twenty years (i.e., exceeded the index every twenty years or so) is suddenly to be expected every five years—and can therefore cause much greater damage.

If insurance companies want to profit from this model in the long term, they will need to keep raising premiums. At some point, many poorer countries will not be able to afford it—even today, some cannot or do not want to pay. The poorest of the poor will have very few options to escape their predicament.

Attribution science may provide one solution. We could begin by calculating how the risk of climate damage has changed in a specific location and to what extent we can attribute this to climate change. This portion of the risk could be covered by an international fund paid into by industrialized countries. It would therefore be worthwhile for insurers to continue doing business in developing countries,* who would continue paying their usual premiums but still receive full protection.

Climate risk insurance is proven to work and is already relieving the burden. However, this has little to do with climate justice—the role of climate change is not taken into account and the richest countries do not systematically partake in the costs. Industrialized countries subsidize insurance policies with development aid sporadically and in individual cases only.

The worst alternative would certainly be to ignore climate damage altogether or to saddle already struggling communities with the cost. If this is the case, there may be only one nonpolitical means of achieving climate justice: lawsuits.

---

* Even now, insurers are only making a profit from many countries because of the millions contributed by countries like Germany and institutions like the World Bank.

If a satisfactory solution cannot be reached at a national or international level, the victims of climate change could play this card. If, for example, entire sections of the population lose their livelihoods because islands have become uninhabitable, they could take the companies and countries with the greatest carbon emissions to national courts or even the European Court of Human Rights.

CHAPTER 9

COUNTRIES AND CORPORATIONS ON TRIAL

I F YOU WERE paying close attention to the British or German newspapers on April 6, 2018, you might have noticed a minor sensation buried in the back pages: the Supreme Court of Colombia had ruled in favor of twenty-five young people who had sued the Colombian state for inaction in the fight against climate change. Aged seven to twenty-six, the plaintiffs argued that the destruction of the rainforest and the associated acceleration in the greenhouse effect would seriously impair their lives and health.

The Colombian rainforest covers an area almost the size of California. However, deforestation has increased significantly in recent times due to agriculture and cattle farming. The plaintiffs made the case that, in standing idly by, the government was violating constitutional rights to life, freedom, and property.[1]

This climate lawsuit was a novelty for Latin America. Very few people expected it to succeed, especially considering the young plaintiffs and the wide-ranging allegations.

And yet the court upheld the complaint and gave the government four months to draft a plan of action to restrict deforestation in the Amazon region.[2]

For decades, energy companies and governments have known about the climatic consequences of burning fossil fuels and the damage they are inflicting on future generations. Nevertheless, very few have rethought or fundamentally altered their business models or policies. And so, in 2017, the planet emitted more carbon dioxide than ever before.[3] The question was, how long would those responsible continue to get away with it? It would only be a matter of time before our children revolted, feeling cheated of their future—or at least that of their children and grandchildren—and attempted to seek justice.

That time starts now.

In 2015, a Dutch citizens' alliance went to court on behalf of future generations, and with success: the District Court of The Hague ordered the governments to do more to protect the climate and to align its self-imposed goals with the findings of the IPCC.[4] In 2015, children and young people in several U.S. states also took their governments to court, demanding that they fulfill their climate goals; most judgments have yet to be announced.* And in India in 2017,

---

* The District Court of King County (Washington) rejected an initial lawsuit by the youth alliance in August 2018; further lawsuits continue in eight other states and at the federal level. Supported by Our Children's Trust, the group has secured at least one success so far: the Trump administration attempted to stop the suit at an early stage, but an appeals court in San Francisco rejected this request in early 2018. (Gustin, "Judge Dismisses Youth Lawsuit.")

a nine-year-old girl sued the government to force it to adhere to the Paris Agreement and emit less carbon dioxide.[5] After all, today's children are the ones who will pay for politicians' failure to act.

These are just some of the climate lawsuits being brought before courts in ever greater numbers, focusing mainly on oil corporations in the U.S. and on governments in Europe. Plaintiffs include the U.S. coastal cities of San Francisco, Oakland, New York, and Baltimore, which are demanding compensation from oil companies like ExxonMobil to help them adapt to rising sea levels with dikes and dams;* ten families from five E.U. countries and from Kenya and Fiji along with a Swedish Indigenous youth organization, who want to force the E.U. to strengthen its climate goals;[6] and Swiss senior citizens demanding that their government do the same.[7] According to one list, there have been 920 climate lawsuits in the U.S. to date,[8] and 269 elsewhere[9] (as of October 2018).

These initial successes show that the basic idea behind the lawsuits works: if governments don't do their job and

---

* The New York and San Francisco/Oakland lawsuits were rejected in 2018 on the grounds that this is the responsibility of Congress, not the courts. These states plan to contest the rulings. (Kusnetz and Hasemyer, "Judge Dismisses Climate Lawsuit"; Hasemyer, "2 City Lawsuits Dismissed.")

In a similar case, Baltimore sued twenty-six fossil energy companies in July 2018. The plaintiffs believe they have a better chance because the case is being heard by a state court, not a federal court. Their case is based on a Climate Central study, which calculates that flooding caused by rising sea levels has already increased by a fifth. (Circuit Court for Baltimore City.)

don't do enough to put a stop to climate change, then courts can remind them of their purpose. Germany would also be a candidate; in summer 2018, Chancellor Angela Merkel announced that her government would unfortunately not achieve its self-imposed goals. She'd had plenty of time, but greenhouse gas emissions had barely dropped during her twelve years in office.[10] Newspaper reports skirted around the issue, with no major discussions of what is, in effect, a scandal. In the same year, Merkel's government lobbied for Miguel Arias Cañete, E.U. Commissioner for Energy and Climate Action, to drop his plan to step up the E.U.'s climate goals.[11]

The response was swift. At the end of October 2018, three farming families joined up with Greenpeace and sued the government for failure to act. According to the indictment, quoted in *Der Spiegel* magazine, the government had "ceased its actions without legal basis and without adequate rationale or justification."[12] Abandoning the 2020 climate goal would impair the fundamental rights of the farmers to "life and health," "freedom of occupation," and "guarantee of property." Among other things, the organic farmers believed they would suffer crop failures caused by extreme weather exacerbated by climate change.

The Changing Nature of Lawsuits

The list of successful climate lawsuits remains fairly short compared to those that have failed or not been permitted to proceed.[13] Most of the successes relate to specific projects—such as airport expansions or new mines or power stations—that lawyers have succeeded in linking to climate change.* These few successes are largely restricted to countries that have contributed less to climate change, such as Colombia. This doesn't mean that these countries should cease climate protection measures; no matter how small, every action makes a difference.

And yet Colombia and the Netherlands are not the U.S. or China. This may be the only reason why the first lawsuits were successful and have attracted little attention. Or perhaps it is also because these lawsuits have suffered from the same thing that has made international climate negotiations so arduous and sluggish: they almost always focus on the future, not the present. They focus on abstract projections of average climatic changes in a country or continent, not on droughts, floods, or storms happening here and now.

Up to now, these lawsuits have only managed to influence the politics of individual countries, not international politics, which is why the companies with the most emissions have not altered their business models or structures.[14] However, these first successes are probably the start of a wave of lawsuits that will have what it takes to shake up the whole world. This is because the nature of the lawsuits has evolved. For a long time, they focused solely on the rights of future generations, on false information issued by corporations, or on governments that did not adhere to their own laws and goals. Now, however, many suits are concentrating

---

* The construction of a new runway in Austria was prevented on the grounds that it would increase air traffic at Vienna's airport and therefore impede the country's national climate protection goals.

on specific damage caused by climate change—damage caused by extreme weather events and rising sea levels. And this is where we come into play.

This approach started in 2009 in Kivalina, a village in the far northwest of Alaska home to four hundred Iñupiat. Kivalina is regularly subjected to severe winter storms. For a long time, it was protected by a shield of sea ice. Thanks to global warming, these ice barriers are melting more and more, allowing winter storms to drench the village in seawater and erode the coastline. As a consequence, buildings were in danger of falling into the Chukchi Sea and the residents were forced to relocate.

And so they sued the people they believed had caused their plight—crude oil corporations like ExxonMobil and coal companies like Peabody Energy—and accused them of conspiracy. According to the plaintiffs, these companies had deliberately deceived the public on the climatic repercussions of burning fossil fuels. In addition to the accusation of misinformation, this case added a new aspect: compensation for the loss of the village. However, the court did not even bother to examine the cause of the damage, but rejected the lawsuit before it reached this stage, stating that responsibility lies with the U.S. Congress, not the courts. It also said that the problem was too abstract and that it was barely possible to determine the reason for Kivalina's demise.[15]

The New York Times wrote that "global warming is a diffuse worldwide phenomenon; a successful public nuisance case requires that defendants' behavior be directly linked to the harm."[16] Event attribution science was still in its infancy at this time.

A damage suit of this kind was also filed in Germany in 2015. The plaintiff was a man who did not speak a word of German and had never been to Germany before his lawsuit was filed with the District Court of Essen in December 2015. Saúl Luciano Lliuya is a farmer from the Peruvian Andes who lives just outside the city of Huaraz in a mountain village 10,171 feet above sea level. From his potato field, he can see the Cordillera Blanca, the "white range." But the mountains are becoming less and less white as time goes by; the snow and glaciers are melting as global warming progresses and the dark rock beneath is showing through.[17]

The meltwater flows into the Palcacocha glacial lake, which is around 3,280 feet above Huaraz and full to bursting. If a large chunk of ice were to break off and fall into the lake, it could create a wave that would flood large parts of the city and surrounding area, including Lliuya's farm. In 1941, a piece of the glacier fell into the lake and the resulting wave killed 1,800 people.[18] The water level is much higher today than it was back then.

This danger hangs over Lliuya's head every day. A dam could protect him and the city, but that would be expensive, and Huaraz either cannot or does not want to pay. Lliuya hit upon the idea that the people responsible for his situation should pay to fix it. In his understanding, these include RWE, an energy company based in Essen, Germany. With the aid of the German environmental organization Germanwatch, he has filed a lawsuit against Europe's biggest emitter of carbon dioxide. According to Lliuya, RWE's emissions have contributed to the melting of the Andean glaciers and helped place his village under threat of flooding from the glacial lake.[19]

This lawsuit has a much more substantial basis than that of the Iñupiat in Alaska. Given that RWE is responsible for almost 0.5 percent of all greenhouse gas emissions in the world, the company would have to cover 0.5 percent of the money that Lliuya's hometown now has to spend to protect against potential flooding. This amounts to €17,000.

While this negligible sum wouldn't exactly be painful for RWE, it could set a precedent if Lliuya were to win—and launch many more climate lawsuits around the world.

For this to happen, it must be proven that the carbon dioxide molecules emitted by RWE have actually contributed to the melting of the glacier above Lliuya's farm. The lawsuit is based on Section 1004 of the German Civil Code, according to which "the owner may require the disturber to remove the interference"[20]—a section that these days is mainly used in neighborly disputes in which causality is easily determined. This is not quite so easy in the case of Lliuya v. RWE AG.

The District Court of Essen rejected the suit in the first instance, stating that a "linear chain of causation" cannot be determined from RWE to the glacier. However, the proceedings are not yet complete, and it will be very interesting to see how things pan out before the Higher Regional Court of Hamm and then, quite possibly, the Federal Court of Justice in Karlsruhe. After all, this damage actually exists; it is stated in the Paris Agreement in black and white, ratified or accepted by every country on Earth. And, unlike with Kivalina, the chain of causation can definitely be traced. With the aid of attribution studies, we can do what was apparently impossible according to the rejection of the Iñupiat lawsuit: calculate the role of climate change in individual extreme events and attribute the climate damage to individual countries or companies. In Lliuya's case, this would involve a lot of work for us scientists because it does not refer to a meteorological event; we would therefore need to realistically simulate both the weather and the behavior of the glacial lake. This isn't impossible, but it is challenging. For the purposes of this book, however, this case merely illustrates nascent court proceedings. Lliuya was the first to face RWE in court, but he won't be the last. Climate change has caused plenty more damage, and gaps in legal arguments can now be filled.

A Global Carbon Inventory

These court proceedings would not have been possible without Richard Heede, a geographer from California. By his own account, Heede has spent fifteen years scouring archives to determine the emissions of individual companies and their legal successors since the start of the Industrial Revolution—and, therefore, how much they have contributed to climate change. He has concluded that just ninety companies have contributed 63 percent of the greenhouse gases emitted worldwide between 1751 and 2010. Half of these emissions were released into the atmosphere only since 1988—after the IPCC was founded and we were all made aware that climate change exists, is a threat, and can be measured.[21]

Heede's carbon inventory can be used to determine the specific contributions of individual companies—and this is where it gets interesting. According to the inventory, Saudi

Arabia's state energy company Saudi Aramco and U.S. oil giants Chevron and ExxonMobil are each responsible for more than 3 percent of humanity's global greenhouse gas emissions since industrialization. The British oil company BP, Russian natural gas giant Gazprom, Royal Dutch Shell, and the National Iranian Oil Company are each responsible for more than 2 percent.

Heede has taken the first step in creating a solid foundation for climate lawsuits against those who have earned money by producing or burning fossil fuels: he has created an inventory of the greenhouse gases emitted by individual companies. The second step—linking this with global warming—was taken by scientists and employees of the NGO Union of Concerned Scientists and their colleagues.[22] To help the second generation of climate lawsuits achieve success, we need to complete the other end of the causality chain: the link between global warming and specific damage caused by extreme weather. Previous lawsuits have neglected this, certainly due in part to a lack of concrete figures. Thanks to attribution science, the figures are now, in principle, available, at least for the biggest fossil fuel producers. And we have everything we need to extend this to other companies; we just need to fit the pieces into the puzzle.

I would not venture to say what effect such climate lawsuits will have in the coming years. But one thing is certain: there will be a shift in the discussion on how rich countries and companies need to assist poorer nations to overcome the consequences of climate change. Those who have neglected to seriously reduce their greenhouse gas emissions should feel less confident about the future.

We Are Not All Equally Guilty

If we haven't already, we now need to ask whether this is actually just. Should courts really take legal steps against corporations for causing or accelerating climate change? Is it really their fault?

Energy companies like ExxonMobil and RWE like to point out that they mine and burn oil, gas, and coal for the benefit of everyone; this is the basis of all prosperity. If anyone is guilty, then we're all guilty, not individual companies.

This argument is misleading. With this attitude, nothing would ever change, and we would provide future generations with a climate far more hostile than the one we have today—perhaps even with extreme weather currently considered impossible.

Placing our trust in politics alone is no longer enough. Even now that the Paris Agreement has been ratified and is binding under international law, most governments do not take their obligations seriously enough. I don't even need to mention Donald Trump's America. Germany, for example, began to clear the Hambach Forest, one of the country's oldest forests, in late summer 2018.* RWE wants to get its hands on the rich seam of brown coal beneath the forest, even though this is no longer necessary to secure Germany's

---

* Following a lawsuit by the BUND environmental organization, the Higher Administrative Court of Münster put a temporary stop to the clearing in October 2018. BUND argued that clearing the forest violated European environmental legislation. It is now expected that RWE will not be able to resume clearing until 2020 at the earliest. (Smith-Spark, "Hambach Forest Clearance Halted.")

electricity supply; renewable energies already cover almost 40 percent of energy consumed.

RWE argues that it is not burning fossil fuels for its own benefit, but for consumers. If they didn't buy electricity generated by coal, no fossil fuels would be burned—so we are all guilty.

It is true that consumers have power, and greater demand for clean energy and vegan food will lead to a wider range of products. But blaming us all for participating only works to a certain extent. Our entire infrastructure is based on fossil energies; even if we make a concerted effort, we cannot lead a normal life without emitting greenhouse gases. Using only solar energy, cycling everywhere, and maintaining a climate-neutral diet is simply not currently possible if you want to participate in society.

Passive houses are now available, but they would definitely exceed my budget, and even these houses are made from materials that generate greenhouse gases during production and transport. Many people are able to cycle to work—I do it every day—but that's only because I can afford to live in Oxford itself and not in the suburbs. When it comes to food, even locally grown organic vegetables must be transported to markets or delivered to the city—generating emissions.

I'm not saying individuals can't achieve anything—they have a great deal of power. But it takes a lot of organization, information, and money for consumers to exercise this power. Corporations can make changes much more easily and effectively by modifying their business models. But that won't happen as long as the old model continues to earn them money and is permitted by law—as long as someone continues to buy their products and there is no risk of being sued for costly damages.

If just one lawsuit against RWE, Chevron, or ExxonMobil were to succeed and actually hurt the company, all other corporations with high greenhouse gas emissions would consider switching to green energy much faster than before.

Green options aren't available for everything, but almost. There are now largely climate-neutral ways of producing cement, but nobody uses them because they are a lot more expensive. The prospect of being sued for lots of money, suffering operational disruption, or falling out of favor with customers could change that.

We also need to ask whether it is possible to determine *in law* that a corporation has acted wrongfully—that it was their duty to protect people against the consequences of climate change, and that they have actively dodged this duty. A lawsuit against the tobacco industry can prove more easily that a company has deliberately caused damage or continued to operate in full knowledge of this damage. While burning fossil fuels does have positive effects—there's no denying that electricity is useful—the positive effects of smoking are extremely limited.

Of course, we have long known about the consequences of burning fossil fuels and releasing greenhouse gases into the atmosphere. Renewable energy now offers a real alternative. And yet it is difficult to prove that a company has caused damage it should not have caused, and has done so deliberately. Many early climate lawsuits were not admitted to court precisely because the question of when corporations received what information and were required to act accordingly is far more complex than in tobacco industry lawsuits, for example.[23] There are still no laws prohibiting greenhouse gas emissions. In contrast to other waste gases, they are often not even regulated; ships are permitted to burn the dirtiest heavy oil without any sanctions. And while regulations are undergoing many changes, it will probably be a long time before bans are actually enacted—if ever. However, such laws may not be necessary for climate lawsuits based on attribution studies to succeed.

Climate Lawsuits of the Future

Jurists are closely following developments in our discipline. Some of them consider attribution studies a key element of future climate lawsuits.[24] The reasons are simple: Our studies enable specific climate damage to be calculated. But beyond this, some jurists believe it is even more crucial that numerous studies already exist showing how much more likely specific extreme events have become because humanity has released greenhouse gases into the atmosphere. The more studies there are—and more are being published every week—the clearer the impact and consequences of climate change become. Not as an average or on a global scale, but at specific times in specific places.

While we have long known that the risks of drought are probably increasing in several semiarid regions, attribution tells us that a drought such as the one that brought Cape Town's water supply to the brink of collapse will occur on average every hundred years today, and around every thirty years in a world that is 2°C (3.6°F) warmer. In a world without climate change, it would have occurred every three hundred years. There is no way for anyone to claim ignorance.

Lawyers Sophie Marjanac and Lindene Patton—with whom I have attended many conferences, led podium discussions, and given talks to students—argue that laws prohibiting greenhouse gas emissions are not necessarily required.[25] The sheer number of studies highlighting the specific consequences of extreme weather ensures that nobody can close their eyes to the reality of climate change. Rather than apportioning blame, jurists focus on gathering facts from which nobody can hide. We now have these facts.

The question is no longer whether courts will use attribution studies, but when. The first such damage suits are already being prepared.

But what form might they take?

Essentially, it all revolves around the share of a drought, flood, or cyclone that can be attributed to a company or country. This means that a company can only be held accountable for the damage linked to climate change. If it is clear that climate change has made a drought 20 percent more intense, then we can translate this increase into the share of damage—either through rough estimates as in the *Heavy Weather* report or with the aid of economic damage functions, empirical equations that convert water levels into dollars, for example.

The role of the company or country must also be clarified in terms of all greenhouse gases released by humanity since industrialization. Emissions can be taken apart like a puzzle and shared among the companies and countries responsible. Calculating company contributions is a Herculean task; it is a little easier for countries because inventories of annual emissions have long been kept for all nations. Reporting these emissions is one of the duties stipulated by the Paris Agreement and its predecessors.

Responsibility for a Heat Wave—in Figures

Through years of detective work, Richard Heede has worked out the numbers for the companies with the greatest carbon dioxide emissions. Together with our Environmental Change Institute in Oxford, the Centre for International Climate and Environmental Research in Oslo has run the figures for the countries committing the greatest climate sins[26] and translated the emissions* of individual countries into shares of global warming.

It has concluded that the main countries or groups of countries responsible for the world becoming 1°C warmer on average since the start of industrialization are the E.U. (17 percent), the U.S. (just under 16 percent), and China (about 11 percent).

But is it really that easy?

The U.S. could argue that the consequences of greenhouse gas emissions were not known at the start of the Industrial Revolution and, therefore, nobody can be held

---

* All greenhouse gases and other industrial emissions were calculated together, including minute particles that are extremely harmful to human health when inhaled but have a cooling effect in the atmosphere.

accountable—or only since we all became aware of the effects. In principle, we became aware of the effects with the work of Swedish physicist and chemist Svante Arrhenius, who caused a furor in specialist circles and was mentioned in the newspapers in 1912.[27] However, the greenhouse effect and its causes were forgotten until the middle of the century, at least among the general public. But nobody has been able to claim ignorance since 1990 at the latest, the year of the first IPCC report. If we count emissions just from 1990 onward, we get a different result: China leads the way with 12 percent, while the U.S. takes second place with 11 percent and the E.U. has just 9 percent.[28]

Critics of this calculation method state—not without reason—that it discriminates against countries such as China and India that have only recently begun to profit from industrialization. The year from which emissions are calculated isn't the only topic of discussion. The type of emissions calculated are also subject to dispute: the main greenhouse gas is carbon dioxide, as this forms the largest share of climate-active emissions and remains in the atmosphere for centuries. If we count just these emissions since industrialization began, we get a different result again: the U.S. is the clear front-runner with 26 percent, followed by the E.U. (23 percent) and China (10 percent).[29] According to this calculation method, the U.S and E.U. alone are responsible for about half of the 1°C rise.

There are arguments for and against all these figures. While they may differ greatly depending on the political, social, or legal interpretation, they all clearly show which countries bear the greatest responsibility for global warming.

My Norwegian colleagues and I have used this work as the basis for the next step: asking whether such calculations are possible for specific extreme events.[30] In short, the answer is yes.

Our example was the Argentinian heat wave of 2013, which was made five times (400 percent) as likely by climate change. We broke this down for individual countries—that is, the extent to which the U.S., China, the E.U., or Japan had made the heat wave more likely with their respective emissions. This form of attribution for individual nations had not been performed before, so I first had to develop the methods.*

The result was that the U.S. and the E.U. had both increased the probability of the Argentinian heat wave by just under 30 percent, China by about 20 percent, followed by India, Indonesia, and Brazil (each around 10 percent), Japan (7 percent), Canada (5 percent), and the remaining industrialized countries, including Australia (7 percent in total).

The article was published in 2017 and cited in legal journals as a specific example of what attribution science can do: complete the chain of evidence from the emissions of individual countries or corporations through to a particular extreme weather event. It was precisely this "linear chain of evidence" that was missing from Kivalina's and Lliuya's lawsuits in the opinion of the court.†

---

* I essentially came up with two ways of performing the statistical calculations. The differences between the two methods are comparatively minor; one is parametric, the other is not.

† There is as yet no chain of evidence in Lliuya's case, but our study on the Argentinian heat wave shows that it is possible.

Admittedly, I am not a lawyer, and my interpretation of this highly complex legal topic comes from conversations with jurists focused on this issue and from reading legal articles. But such evidentiary value should not be all that unusual in court proceedings. For some time, courts have been dispensing justice on the basis of proportional liability and forms of evidence that take probabilities into account. Past cases have covered similar terrain, such as damage suits from uranium miners who were inadequately protected and later diagnosed with cancer.

Attribution science also provides another point of reference in public life. The New Zealand Treasury commissioned climate scientists David Frame and colleagues to calculate how much climate change has cost New Zealand in the last ten years. He and his colleagues used attribution studies to estimate climate change's share in the most expensive extreme events between 2007 and 2017. The Treasury adopted the conservative assessment that climate change has cost the country NZ$120 million due to the increased risk of flooding and NZ$720 million through droughts.[31] If attribution studies are good enough for governments, why shouldn't a court base its judgments on such calculations?

There are still some obstacles to overcome. For example, how will courts deal with uncertainties that are totally normal in science but could impede a legal argument?* And how

---

* Probabilities and how they are altered by climate change—known to scientists as the "risk ratio"—are fraught with uncertainty. This is a normal part of science, reflecting the fact that perfect measurements and perfect models do not and cannot exist. Every scientific discipline works with incomplete data and models based on assumptions. These assumptions are

will they cope with the fact that the same extreme weather can be defined in very different ways, and that plaintiffs and defendants may each present attribution studies with different but correct results?*

What Scientists May and May Not Do

Scientists are sometimes accused of allowing ourselves to be exploited by environmental activists. Some scientists frown on climate lawsuits and attempts to identify guilty parties, believing that science should be devoted solely to the pursuit of knowledge. This is, of course, nonsense.

We too are only human and our political convictions and values naturally influence what and how we research. What is crucial is that scientists remain impartial; this is why we

---

often justified and sound, and uncertainties can be quantified. However, this is often the most complex part of a study, and judges are not scientists, with good reason. And while it is relatively easy for scientists to agree that climate change has made a specific event more likely, or to determine the scale of the event, it will be much more difficult to provide exact numbers—although not for all events.

* For example, recall that a heat wave can be defined using a specific temperature value, but also in terms of heat stress. The probability of the heat wave changes depending on the perspective. This is not a problem if a study is being used to plan adaptations—heat stress is used for health prevention, temperatures for planting periods. If, however, you wish to sue someone for damages, these various interpretations may cause problems. While we still do not have a single, correct method of attributing extreme events, there are now so many good studies that it would at least be difficult to present arguments based on anything other than evidence that meets scientific standards. The crucial question will be what makes a meaningful definition of an extreme event, and answering this will be aided by the existence of multiple studies with corresponding examples.

publish all our work under peer review, disclose our funding sources, and do not take on projects if the financiers want to keep the results private. It is important that we retain all rights to the data we produce. Our insistence on this at World Weather Attribution has led to a few abortive projects over the years, for example with infrastructure providers who wanted to identify facility locations where the risk of extreme weather has particularly increased, but who wanted to keep the results private.

Courts cannot replace politics, of course, and they shouldn't. But with one or two spectacular cases or several smaller successes, climate lawsuits could help to make some changes, particularly in parts of society that are rarely convinced by good arguments: energy companies that want to cling onto the past, but also governments who consider their sole mandate to be making policies "for business" or for the supposed good of the economy—which would often be better served with sustainable, long-term economic frameworks.

Courts have an even more important role: ensuring justice for all those who do not have the opportunity to make their voices heard, such as future generations; our children are not allowed to vote, but they will have to live with the damage wrought by climate change.

In Colombia, the children's voices were heard. If they had demanded millions of dollars in damages from ExxonMobil in the U.S. rather than suing a relatively poor country with a minor role on the world stage, they would certainly have made headlines worldwide. If attribution studies can help achieve greater justice in the world, then I am happy to be accused of playing the blame game.

CHAPTER 10

CLIMATE CHANGE IN EVERYDAY LIFE

Seeing the Weather From
a New Perspective

I N SUMMER 2018, many people began to ask questions. Northern Europe had been sweltering for months, and other parts of the northern hemisphere had also experienced extreme heat. Brits and Germans basked in the sun and the warm air, hotels enjoyed an upswing in trade, and beer sales went through the roof. And yet many people couldn't quite bring themselves to trust this unusually long, unusually hot, and unusually dry season.

That summer, I was astonished to find that wherever I went, people were starting to talk about climate change. In cafés and bars, on trains and airplanes, in offices and on the streets. If you listened hard enough, you could catch snippets of the same questions: Is this normal? Or is it climate change?

Journalists from all over Europe wanted answers too. Geert Jan and I found ourselves giving interview after interview, comparing our scientific findings with the events playing out around Europe.

Everyone could see what was happening. Taking the train across Germany, I looked out of the window and saw a Mediterranean landscape: dried-up, bleached-out fields, hardened and cracked ground, lakes reduced to bathtubs. The River Elbe held so little water that you could even see some of the "hunger stones" used to mark low water levels. Unlike high-water marks, these stones usually remain hidden for many years, sometimes for decades or even centuries.

Just over the border from Germany in the Czech Republic, a hunger stone was revealed in the Elbe riverbed on the west bank of the town of Děčín. It bore these words: "If you see me, weep." The message had probably been carved into the basalt in the nineteenth century to commemorate the arid years in which farmers lost their crops and the people starved. It was also engraved with dates of extreme summers: 1868, 1842, 1800, 1790, 1746, 1616. The inscriptions for 1473 and 1417 had almost worn away.[1]

Other parts of Europe were also exceptionally hot, including the Netherlands and Great Britain. The latter experienced its warmest summer ever recorded, on a par with the legendary summer of 1976.

When the end of July brought forest fires to Sweden, Greece, and parts of Great Britain and Germany, we were contacted by increasing numbers of journalists looking for more than just a one-sentence statement that more heat waves are to be expected in a warming climate. We decided to conduct an attribution study—and quickly.*

We concentrated on northern Europe because there was plenty of data available and we are familiar with European weather. It would have taken a long time to incorporate heat waves in the rest of the world, and we wanted to deliver the facts as quickly as possible. In northern Europe, temperatures particularly deviated from the long-standing average between May and July, and above all in Scandinavia, Great Britain, and the Netherlands. This time, however, we did not want to study entire countries as we had often done before; the national average says little about the heat experienced by the residents of Kiel or Utrecht, Dublin or Linköping, Oslo or Copenhagen, or Sodankylä and Jokioinen in Finland. We therefore decided to focus on these particular cities and districts, for which long series of weather data are available, some dating all the way back to 1874. We wanted to find out just how extreme the heat wave really was in these areas and—the question that seemed to be on everyone's mind—determine the impact of climate change.

One of the first things we discovered was that this heat wave was not all that unusual (at least when we conducted our study in late July; some of these cities experienced even higher temperatures in August). Our statistical calculations

---

* In some parts of Europe, the heat was paired with aridity. It is often this combination that determines how people experience a heat wave. It is highly debatable whether climate change has made this combination of heat and aridity more likely. Another attribution study would be required on this topic; our study focused solely on the heat wave, not on the drought.

showed that a heat wave of this magnitude now occurs in Utrecht every five years, and every eight years in Dublin and Oslo.[2]

But how could that be? Oslo was experiencing July heat like never before. Did this mean that, while temperature records were broken, the heat wave was in no way a rare event?

This apparent contradiction is easily resolved. To understand it, we need to know that climate change has increased temperatures in almost all global regions. While people may be experiencing heat they have never experienced before, this is now totally normal from a statistical perspective because climate change has moved the yardstick. If we had the same climate system as, say, 250 years ago, three consecutive days of 31.2°C (88.2°F) in Oslo (as in summer 2018) would certainly have been an extreme event—but in today's climatic conditions, this is no longer very rare.

Oslo's summer of 2018 was a prime example of what climate change means for people's everyday lives: these record temperatures are the "new normal."

Next we determined the extent to which climate change had increased the probability of these July heat waves. Our computer simulations showed that it had made the heat wave twice as likely in Dublin, three times as likely in Oslo, and five times as likely in Copenhagen. Climate change was making its presence felt, just as it was in Kiel, Utrecht, and Linköping. All returned similar numbers—as I said, this is the new normal.

The 2018 Heat Wave:
Global Media Presence

We began the study on Tuesday and published our results on Friday.* None of our team expected the study to have such an explosive—and global—impact. By the following Tuesday morning, just four days after we presented the study, it had been reported by over 2,500 print media outlets alone, not to mention radio and television. And not just in Europe, but worldwide, including the BBC, *Scientific American* magazine, and the Chinese portal Xinhuanet. Harvey was our only other study to receive such a huge response.

When we launched World Weather Attribution, we never imagined attracting so much attention. Back in 2014, most people in Europe and the U.S. still saw climate change as something happening elsewhere in the world—if they believed it was happening at all—or as something that might be a problem for our children or grandchildren but had nothing to do with us.

Considering the number of reports in 2018, this attitude appears to have changed. Climate change seems to have taken root in people's minds, or they have at least become more open to the concept. This is certainly not just due to the 170 attribution studies we and our colleagues around the world have conducted since 2004 on more than 190 extreme weather events, but I'd say they've played their part.

---

* Our previous work enabled us to proceed quickly: the 2018 study was essentially a repeat of our study on the Mediterranean region from summer 2017.

A survey from August 2018 showed that, having experienced a sweltering summer, 72 percent of Brits were concerned about the consequences of climate change.[3]

Throughout August, Geert Jan, Robert Vautard, and I gave interviews on the heat wave practically every day. Suddenly, everyone wanted to discuss the effect climate change was having on their doorstep—even though the study focused on northern Europe and did not cover North America. Nor did it cover Japan, where temperatures exceeding 41°C (106°F) had killed dozens of people and hospitalized thousands more.[4]

Media interest in our study was unbelievable—but only for the first part of the study, which stated that climate change had made the heat wave twice as likely in Dublin, three times as likely in Utrecht, and five times as likely in Copenhagen. But at least we achieved one of our goals: to drag climate change out of the future and into the present.[*]

Essentially, the examples in the study confirm what the global climate reports say: more greenhouse gases, higher temperatures, more heat waves. The new normal. I cannot stress enough how significant it is if the theory in the

---

* We are in fact becoming victims of our own success. It is fantastic to be able to present our work to the world. But the media now expect to be provided with real-time information about extreme weather events everywhere, as though we were a weather service instead of a handful of scientists who are not paid to do any of this rapid attribution work, much less attribute every single heat wave to climate change. The time may have come for some of this work to be taken on by people and institutions who have the facilities to offer operational weather services.

textbooks actually translates to the real world and can be shown or quantified. The demand for these figures shows how important they are.

However, it is not enough to simply promote climate change as a topic and to confirm what was already expected.

What Even Is Normal Now?

I have yet to mention two of the seven locations from our study—Sodankylä and Jokioinen in the far north and south of Finland—and you won't find them in any of the media reports. And yet these cities may actually teach us more about what climate change is and how it manifests in our weather than findings on average temperature behavior. We had good weather data for these locations, which also experienced high temperatures in summer 2018. Very high indeed: located north of the Arctic Circle, Sodankylä hit 31.9°C (89.4°F), several degrees higher than any July temperatures previously recorded.

The generally high variability of summer temperatures here and the fact that 2018 was so unusual make it impossible to determine a robust statistical average. Given the clear margin between these and other temperatures previously measured (in contrast to Oslo, where the record was narrowly broken), we can only estimate. We must therefore use statistical models and proceed as though we had much longer measurement series than just for the last 110 years. This is where things get tricky. To generate these fictional measurements, we need to use several models to obtain reasonably informative data; this also means that our result

won't be just one number, but a very broad spectrum. Therefore, we only know the upper and lower limits of how rare
a heat wave is in Sodankylä where, according to these figures, a heat wave of this magnitude occurs at least every
90 years.* In southerly Jokioinen, it is expected to occur at
least every 140 years. Unlike in Copenhagen or Dublin, for
example, these heat waves are actually extreme and are not
categorized as the "new normal"—instead, they form a new
category.

We experienced even more difficulties when attempt-
ing to attribute this heat wave to climate change. Finnish
summers are changeable: sometimes the temperature drops
to freezing, other times it exceeds 20°C (68°F). This variability makes it almost impossible for us to calculate with
confidence which temperatures are possible in worlds with
and without climate change; almost anything is possible in
either world. The only thing we can say for certain is that
climate change has made the heat waves more likely.

Although we are talking about one and the same heat
wave, triggered by a high-pressure area over Scandinavia,
the story is very different for Oslo and Sodankylä. While
summers in Oslo tend to stay more or less the same, making it even more obvious when the temperature fluctuates
(potentially aided by climate change), Sodankylä's summers
are so fickle that the signs of climate change really need to
stand out.

---

* The lower bound is more certain: it depends much less on the choice of
  methodology, is influenced less by very extreme events, and is better supported with data.

Why is this significant? While climate change in Finland is important, of course, this is about what is actually normal in the modern era of climate change. A "normal" Finnish summer is defined fairly vaguely and covers a wide spectrum of temperatures, while summers in London and Utrecht are fairly uniform with narrow boundaries. So while a much hotter summer in the south of northern Europe would be a good example of how climate change is affecting our everyday lives, a hotter summer in the north of northern Europe isn't—or at least is a less useful example. The more boring the weather, the easier it is to pin down climate change.

Perhaps the most important task of our World Weather Attribution team is, therefore, to figure out which weather is actually still normal in the world in which we live. In northern Europe, it is normal for summers to vacillate between hot and cold, while summer is always hot in southern Europe. It might sound mundane, but we need to know what weather is actually possible before asking how it has been altered by climate change.

We must also remember that climate change may not be the reason why an extreme event has become more likely. There may be other reasons for a disaster—such as deforestation, poor urban planning, or simply an unusually warm El Niño year or the chaotic variability of the weather itself. Even a hot summer doesn't change the fact that weather and climate are not the same thing and that climate change cannot be seen in every weather event. Of the 190 cases of extreme weather we have studied so far—most of them heat waves, droughts, extreme rainfall, and floods—climate change has made around two-thirds more intense or more likely.[5] And those are just the events that have been studied; many more extreme weather events have occurred over the past years for which the role of climate change has not yet been researched.

Climate Change Has Arrived in Everyday Life

Admittedly, this is all fairly complicated. And yet attribution studies are already helping to provide clarity and to pinpoint the effects of climate change in everyday life. The real breakthrough will only come when people think about climate change all the time—not just when experiencing a lengthy heat wave, but in the midst of a cold, gray winter when they find themselves longing for hot weather.

Most people in central Europe are slowly beginning to look at the weather in a new way. We didn't need to worry about it all that much before, unless deciding whether to carry an umbrella or wear a jacket. If the German Meteorological Service (DWD) issues a storm warning, the only people who'll take notice are likely to be festival organizers and rail companies. Most people will barely alter their plans, and that's if they pay any attention at all. The situation is very different in many areas of the U.S. It isn't that Germans don't trust their weather service (the reverse is true, if anything) or that they aren't as smart. But in the American Midwest, everyone knows that the weather can be dangerous and that truly terrible weather is more than just an inconvenience—it can kill. If a tornado warning is issued, they don't go out in their cars; they stay home, make sure they're safe, and maybe even head down to their storm shelter.

This isn't to say that climate change will transform central Europe so drastically overnight that the weather will start killing loads of people on a regular basis; the 2003 heat wave was an exception. But climate change is making noticeable changes to our everyday lives via its manifestation, extreme weather.

Take floods, for example: when rivers burst their banks in southern England in January 2017, it was in part due to climate change, which had made the floods more likely (although only a little). These floods can now be expected every seventy years, instead of every hundred. Most people will still only experience them once in their lifetime. And they wouldn't have been fundamentally different without climate change. While such gradual changes barely register in our everyday lives, at least directly, they are definitely picked up by insurance companies: reliable evidence of even a slight increase in water levels and the probability of flooding could dramatically reduce real estate values if insurance companies alter the way they calculate risks and houses are suddenly deemed to be in high-risk zones. Thanks to climate change, some houses in Great Britain's Thames river valley are now situated in a flood area. It may not affect that many houses, but if one of them is yours, it will definitely be important.[6]

Then we come to heat waves. Climate change has made Mediterranean heat waves around one hundred times as likely. In other words, their probability has increased by a whopping 9,900 percent. This will seriously affect the people living in the region—especially elderly people who are already weak and susceptible to heat stress.

If heat and aridity team up as they did in summer 2018 (in which climate change played a comprehensive role), the consequences for humanity can be severe—including forest fires and withered fields. In such cases, it is extremely important to know whether a summer like 2018 will occur more often. The government might be able to spend millions supporting farmers for one summer, but they can't do it every year. If in doubt, they will have to consider other instruments such as weather insurance.

Attribution studies are also important because most of us only realize how vulnerable we are when caught off guard by storms, droughts, or floods—and by then it's too late. Or when we avoid disaster by the skin of our teeth: when the rain returns and fills our reservoirs, or a dam holds because the rain has eased. Attribution studies provide valuable assistance in cases like these, determining whether weather events previously so unlikely that they could be ignored are still sufficiently infrequent that we do not require extensive adaptation measures.

Floods: Annoyance or Existential Threat?

Finally, I would like to demonstrate the unexpected ways in which climate change can infiltrate our everyday lives. Most Brits (along with countless others around the world) share a love of soccer. Every team is tied to a specific sports ground, some of which are located near rivers or other bodies of water. Even a slight increase in flood risk can affect an entire league—particularly in winter, the main season for British soccer. Not necessarily for the major leagues, whose teams have plenty of money and large, weatherproof stadiums, but the lower leagues, which might attract less attention, but have far more players.[7]

For nearly fifty days in the 2015/16 season, the stadium of League Two team Carlisle United FC was forced to close after Storm Desmond caused major flooding in the area. You might recall that Desmond was one of the first extreme weather events we were able to attribute to climate change.[8] Fifty days without a stadium in the middle of the season caused real problems for Carlisle United, mainly financial. Even a few days lost to climate change can be a few too many. The smaller the club and the more it depends on income from home matches, the more it will be affected by such changes in the weather. What might be "nuisance flooding" to some could change someone else's entire life, even that of a local soccer team.

You don't need to study physics to understand the results of studies asking whether and how climate change is making extreme weather more likely. Admittedly the calculations are a little complex, but if you consider them important, you'll figure them out. Essentially, journalists, newspaper readers, mayors, scientists, and NGO employees need to ask themselves just four questions to classify the role of climate change in their everyday lives or plans:

- To which extreme event(s) am I vulnerable?
- How extreme is this event?

- Has the probability of the event changed and, if so, how much?
- How great are the uncertainties?

Most of us probably don't consider which weather could cause us actual harm. I've only thought about it once: when buying a house, I had to inform the insurance company of the flood risk. My house is on a hill, so it didn't take me long. Most people probably won't think about it until there's water in their family room, trees are falling on cars, or elderly relatives are admitted to hospital with dehydration.

But the more attribution studies there are, the more newspaper articles we will see about people being confronted with these four questions. If we see extremes becoming the new normal for other people, we might start to wonder what extreme events we could experience ourselves.

Day 52

While the people of Houston and the surrounding area spend mid-October 2017 clearing the wreckage and getting their homes back in order, another hurricane is forming over the Atlantic. It is unusual to see a hurricane so late in the year, but Ophelia is the perfect conclusion to a crazy hurricane season with ten named storms in the Atlantic; this has not happened since 1893.

Hurricane Ophelia developed 186 miles southeast of the Azores. But this time—and this is a lot more unusual—it isn't traveling west to America. Instead, it's heading north—over the Azores, where it fells a few trees and floods the shores, and on to Portugal, where its violent winds stoke up the forest fires.* It then moves on to Ireland, where winds of up to 118 miles per hour rip roofs off houses and snap trees. It will later transpire that this is the first Category 3 hurricane of the satellite era to move so far east. This is also partly due to unusually high ocean temperatures.[9]

On October 16, the storm moves over Scotland and even stretches as far down as southern England. Ophelia may no longer be categorized as a hurricane, but its winds are still powerful enough to rip tiles off roofs, snap transmission towers, and uproot trees.

It carries souvenirs of its journey: a huge, swirling cloud of particles from Portugal's forest fires and the Saharan dust it sucked up after forming in West Africa. The dust scatters the sunlight, which does not diffract as usual; behind this hazy veil, morning dawns on October 16 with an unusual color in England's skies. From my office window, I can see the hurricane's final legacy: an apocalyptic, yellow-red sky.

At that moment, I feel a little regret that we are not yet able to quickly attribute hurricanes to climate change; their physics are complex and climate models still find them tricky. So far, we have only been able to attribute their side effects, such as the extreme rainfall over Houston, but we are working hard to determine humanity's impact on the strength of storms. Ultimately, we want to know whether

---

* In summer 2017, aridity and extreme heat were a particular problem on the Iberian Peninsula, a problem that clearly bore the hallmarks of climate change. (Kew et al., "The Exceptional Summer Heat Wave.")

Harvey, Irma, and Ophelia are the heralds of a new era. We know that the increased rainfall of these hurricanes as well as less extreme events are now more likely—but we don't just want to identify the "new normal." It's time to tackle the next question: What is the "new extreme"?

EPILOGUE

W E ARE THE ones who will decide whether we get the weather we want in the future—particularly for those members of our global community who cannot sufficiently protect themselves against its anger. With every carbon dioxide molecule emitted by a coal-fired power plant, container ship, or minivan, we alter droughts, floods, and hurricanes. Every time we vote, pick an electricity provider, make financial investments, choose a mode of transport, or set the menu for our next family get-together, we decide just how angry the weather will get.

Attribution science allows us not only to look to the past, but also to say what weather awaits us in the future depending on our actions. With the methods developed in our discipline, we can simulate the weather in a world that is 1.5°C, 2°C, or 3°C (2.7°F, 3.6°F, or 5.4°F) hotter.

The Lucifer heat wave that transformed the Mediterranean into a veritable furnace in summer 2017 would have been extremely rare in a world without climate change; today, however, we can expect a summer like that almost every eight years. If the global temperature were to increase by another 0.5°C, making it 1.5°C warmer overall, a heat wave of this magnitude would occur every four years. If it were to rise by 2°C, we would experience these extreme temperatures almost every second summer—and those are the most conservative estimates from all our model calculations. With a 3°C increase, most summers would be even hotter; 2017 would feel cool in comparison.

But 3°C? Isn't that totally unrealistic? After all, didn't the global community agree to restrict global warming to 2°C, even to 1.5°C if possible, when ratifying the Paris Agreement? No. From today's perspective this is not remotely unrealistic; it is precisely the scenario we are heading toward. In the world our children and grandchildren will most likely inherit, summers like those of 2017 in the Mediterranean and 2018 in northern Europe will be exceptions—they will be colder than usual.

We find ourselves on the threshold of a new age of extreme weather. The good news is that we can still do something about it. At least, we can prevent a 3°C rise and the weather it will bring, weather that would give parts of the planet a whole new look—and not a very pretty one.

If this were only about the current generation of Europeans, then you could say we're getting the weather we deserve. Sadly, it's not that simple. While we are faring pretty well and most of us enjoy, rather than tolerate, the warm summers, our children and grandchildren will pay for the havoc we are wreaking. And we are indeed still wreaking it; thirty years after the IPCC was founded, there is still no sign of a reverse in the trend. In 2018, more greenhouse gases were emitted worldwide than ever before.[1]

Most people on Earth get weather they don't deserve, people who have benefited very little or not at all from the burning of fossil fuels. Thailand, for example, will have to reckon with more frequent extreme rain such as that of 2010, which submerged large swaths of the country and killed more than 250 people. Even today, climate change has made such deluges twice as likely; in a 2°C world, they would occur even more often.[2]

In northwest India, the probability of heat waves like that of 2015 has not yet increased. But in a 1.5°C world, these heat waves will become twice as likely. In a 2°C world, they will be a full ten times as likely; they will become normal.

Of course, climate change is not making everything worse everywhere; there are occasional glimmers of hope. Peru, for example, will be increasingly spared from cold spells like that of 2013, which killed over five hundred people. However, the country is facing other problems: melting glaciers are threatening the homes of people like Saúl Lliuya, the farmer who took RWE to court.

The case of Lliuya v. RWE AG shows that even a Peruvian farmer no longer has to stand by and watch, powerless, as climate injustice reigns. Attribution transforms climate injustice from a theory calculated in models into a real-world concept.

These developments may finally tip the scales of the global energy system—at least according to the late Elinor Ostrom, an American political scientist and the only woman to have ever won the Nobel Memorial Prize in Economic Sciences. She believed that such transformations happen when groups with little access to power become aware of the power they do have and the actions they can take, and express these through social movements, strikes, and protest marches, by voting, or even through lawsuits. There is plenty we can do to make countries and companies accept and tackle the challenges of climate change. Every single one of us can help decide whether global warming increases by 2°C, 3°C, or 4°C (3.6°F, 5.4°F, or 7.2°F). This is both an opportunity and a responsibility.

Those in the global North have more responsibility, of course; unlike many of our fellow humans, we do not have to spend every day figuring out how to survive. We have the freedom to take action. When I read a newspaper article about the repercussions of climate change or discuss it with someone I know, the tone generally veers between panic (the world is going to end, this is just the start, it will be even worse than we think) and total apathy (corporations will always have the upper hand, people only think about themselves, politicians are all talk and no action). These feelings are all too human. Without numbers, the limits of responsibility become blurred; everything is either terrible or meaningless. Does it make any difference whether I fly to Paris or catch the Eurostar? Who can tell whether and to what extent the things I do today will affect the future?

Science is not a universal remedy for panic and ignorance, and attribution science alone will not save the world. It is, however, an effective tool to provide guidance and show whether climate change has fundamentally changed the rules of specific weather events, played only a minor role, or been wrongfully blamed.

Scientific methods provide a robust bulwark whenever interest groups circulate "fake news," politicians try to downplay the consequences of environmental pollution, or climate activists paint an apocalyptic picture. Well-founded numbers enable us to identify those actually responsible and expose attempts to conceal or even deny uncomfortable facts. They also give us the tools we need to better protect ourselves against the consequences of extreme weather in the future.

Attribution science is only just getting started. Many things that will be possible tomorrow are still pie in the sky today. But the foundations have been laid. This book will have fulfilled its purpose if it increases awareness of the impact climate change is having today and how this is reflected in our weather—and that this is neither a fabrication nor a catastrophe that we are powerless to stop.

ACKNOWLEDGMENTS

I AM DELIGHTED AND grateful to be part of the international community of climate scientists, and to have spent the last few years working on fascinating projects with many different people. I have learned something from every one of them, but nobody has taught me as much as Geert Jan van Oldenborgh. For that, I thank him from the bottom of my heart; without him, there would be no World Weather Attribution, and my career would certainly have taken a different path.

For their criticism, encouragement, support, and ideas, I would particularly like to thank Myles Allen, Emily Boyd, Heidi Cullen, Jan Fuglestvedt, Luke Harrington, Gabi Hegerl, Rachel James, Richard Jones, Sarah Kew, Sjoukje Philip, Roop Singh, Sebastian Sippel, Sarah Sparrow, Claudia Tebaldi, Robert Vautard, and David Wallom, as well as the many citizen scientists who have facilitated our model simulations for over ten years through climate*prediction*.net.

Writing a book is stressful, but it was more enjoyable than expected, and for that I thank my collaborator, Benjamin von Brackel, for turning my brain dumps into prose and making science even more exciting with an array of new suggestions and ideas.

Thank you to Kristin Rotter for putting up with the clashes of personality and experience between Benjamin and myself, which certainly proved enriching. If it weren't for Kristin, I would never have had the idea to write a book, and I am grateful for this idea; working with you has been a joy. Huge thanks also to Dunja Reulein for her fantastic editing.

This book would not exist without Johannes and Alexander Otto. Thank you, Johannes, for all those evenings when you lay next to me and read, quiet as a mouse, as I wrote and edited. Thank you, Alek, for being my first reader, for having an opinion on everything I needed, but above all for constantly encouraging me to write a book, be a boss, and simply to be human.

Extra-special thanks to Peter Walton, Matt Brown, and Karsten Haustein—for many things (like keeping me sane . . .), but mostly for letting me be one of you.

EDITORIAL NOTE

*A*ngry *Weather* is not a scientific discourse on the development of attribution science over the last fifteen years. It therefore does not mention all the key scientists who have helped shape the discipline during this time, nor does it cite all significant publications. Those who are given a voice or whose work is cited have been chosen because they provide examples for the points I wish to illustrate. If one of my colleagues had written this book, they would have focused on different aspects, cited different literature, and featured different people. Sources are provided for all the facts in the book, but the interpretation of these sources is mine and not the only possible interpretation. Much of the story about Hurricane Harvey comes from my memories and will certainly not match my colleagues' recollections word for word. The book reflects my worldview as a proud member of the "footloose elite" and citizen of nowhere that British politicians are so fond of disparaging. My opinions in this book are my own and do not necessarily reflect the stance of World Weather Attribution.

NOTES

Preface

1. Franta, "Early Oil Industry Knowledge of CO_2 and Global Warming."

Prologue: The New Weather

1. Lascaris, "Are Irma-Like Super Storms the 'New Normal'?"

2. Blake and Zelinsky, *National Hurricane Center Tropical Cyclone Report: Hurricane Harvey.*

3. Hauser, "Hurricane Harvey Strengthens and Heads for Texas."

Chapter 1: Cause and Effect: How We Created Our Weather

1. Schaller et al., "Human Influence on Climate in the 2014 Southern England Winter Floods."

2. Schaller et al., "The Heavy Precipitation Event of May–June 2013 in the Upper Danube and Elbe Basins."

3. Timbal, Arblaster, and Power, "Attribution of the Late-Twentieth-Century Rainfall Decline in Southwest Australia."

4. Sippel and Otto, "Beyond Climatological Extremes."

5. Gibbons, "Harvey's Intensity and Rainfall Potential Tied to Global Warming."

6. Ellen, "Fox News' *Outnumbered* Ignores Impact of Climate Change on Hurricane Harvey's Epic Intensity."

7. Van der Wiel et al., "Rapid Attribution of the August 2016 Flood-Inducing Extreme Precipitation in South Louisiana to Climate Change."

8. Otto, "Link Between Hurricane Harvey and Climate Change Is Unclear."

Chapter 2: Sowing the Seeds of Doubt: Climate Change Deniers

1. Supreme Court of the State of New York County of New York: People of the State of New York v. Exxon Mobil Corporation.

2. Schwartz, "New York Sues Exxon Mobil."

3. Bussewitz, "New York Says Exxon Misled Investors."

4. Schwartz, "New York Sues Exxon Mobil."

5. Supran and Oreskes, "Assessing ExxonMobil's Climate Change Communications"; Union of Concerned Scientists, *Smoke, Mirrors, and Hot Air*.

6. Climatefiles, "1982 Exxon Presentation on 'CO_2 Greenhouse Effect' and Exxon Climate Modeling."

7. Supran and Oreskes, "Assessing ExxonMobil's Climate Change Communications."

8. Supran and Oreskes, "Assessing ExxonMobil's Climate Change Communications."

9. Jacques, Dunlap, and Freeman, "The Organisation of Denial."

10. Competitive Enterprise Institute, "Global Warming—'Energy,'" YouTube video.

11. Littlemore, "Heartland Insider Exposes Institute's Budget and Strategy."

12. Oreskes and Conway, *Merchants of Doubt*.

13. Brulle, "Institutionalizing Delay."

14. Jacques, Dunlap, and Freeman, "The Organisation of Denial."

15. Lewandowsky and Oberauer, "Motivated Rejection of Science."

16. Jacques, Dunlap, and Freeman, "The Organisation of Denial."

17. Media Research Center, "Has CNN Warped Meteorologist Chad Myers' View?"

18. Sweney, "BBC Radio 4 Broke Accuracy Rules."

19. Vidal, "Revealed: How Oil Giant Influenced Bush."

20. Krugman, "Enemy of the Planet."

21. International Energy Agency, *World Energy Outlook 2012*.

22. United Nations Framework Convention on Climate Change, *Adoption of the Paris Agreement*.

23. EIKE, "Marc Morano," YouTube video.

24. *Klimaretter.Info*, "Rechter Unions-Flügel folgt Trump."

25. Fischedick, Görner, and Thomeczek, *CO2: Abtrennung, Speicherung, Nutzung*, 823.

26. Global Carbon Project, *Global Carbon Budget*.

27. Samenow, "60 Inches of Rain Fell From Hurricane Harvey."

28. Chappell, "National Weather Service Adds New Colors."

29. Gilmer, "Proximity Counts."

30. ExxonMobil, "ExxonMobil Allocates $500,000 for Gulf Coast Community Hurricane Relief Efforts."

31. Grande et al., "Heartland Institute Experts Comment on Hurricane Harvey."

32. Kalhoefer, "So Far, ABC and NBC Are Failing to Note the Link Between Harvey and Climate Change."

33. Mann, "It's a Fact: Climate Change Made Hurricane Harvey More Deadly."

34. Rice, "Harvey to Be Costliest Natural Disaster in U.S. History."

Chapter 3: Revolution in Climate Science: Turning the Field on Its Head

1. Allen, "Liability for Climate Change."

2. Dole et al., "Was There a Basis for Anticipating the 2010 Russian Heat Wave?"

3. Keim, "Russian Heat Wave Statistically Linked to Climate Change."

4. Otto et al., "Climate Change Increases the Probability of Heavy Rains."

5. National Academies of Sciences, Engineering, and Medicine, *Attribution of Extreme Weather Events*.

6. Fountain, "Looking, Quickly, for the Fingerprints of Climate Change."

7. Davidson Sorkin, "What Has Hurricane Harvey Taught Donald Trump in Texas?"

8. Boburg and Reinhard, "Houston's 'Wild West' Growth."

Chapter 4: The Human Factor: Calculating the Influence of Climate Change on the Weather

1. Old Weather, https://www.oldweather.org/.

2. School of Geography and the Environment, University of Oxford, "About the Radcliffe Meteorological Station's Records."

3. Schaller et al., "Human Influence on Climate in the 2014 Southern England Winter Floods."

4. Vautard et al., "Human Contribution to the Record-Breaking July 2019 Heat Wave."

5. Carbon Brief, "Q&A: How Do Climate Models Work?"

6. Box and Draper, *Empirical Model-Building and Response Surfaces.*

Chapter 5: Heat Waves, Downpours, and More: The Role of Climate Change in the Weather

1. Pierre-Louis, "Climate Change to Blame for Deaths in 2003 Heat Wave."

2. Robine et al., *Report on Excess Mortality in Europe During Summer 2003.*

3. Di Liberto, "India Heat Wave Kills Thousands."

4. Massey et al., "Have the Odds of Warm November Temperatures and of Cold December Temperatures in Central England Changed?"

5. Subramanian, "In Georgia's Peach Orchards, Warm Winters Raise Specter of Climate Change."

6. Otto et al., "Factors Other Than Climate Change, Main Drivers of 2014/15 Water Shortage in Southeast Brazil."

7. Schaller et al., "The Heavy Precipitation Event of May–June 2013 in the Upper Danube and Elbe Basins."

8. Van Oldenborgh et al., "Extreme Heat in India and Anthropogenic Climate Change."

9. Amadeo, "Hurricane Irma Facts, Damage, and Costs."

Chapter 6: Ignore Climate Change and Suffer Its Wrath

1. Amadeo, "Hurricane Harvey Facts, Damage and Costs."

2. Collier and Satija, "A Year Before Harvey, Houston-Area Flood Control Chief Saw No 'Looming Issues.'"

3. Coy and Flavelle, "Harvey Wasn't Just Bad Weather."

4. Wallace et al., "How One Houston Suburb Ended Up in a Reservoir."

5. Sims, "The U.S. Flooded One of Houston's Richest Neighborhoods to Save Everyone Else."

6. Boburg and Reinhard, "Houston's 'Wild West' Growth."

7. Boburg and Reinhard, "Houston's 'Wild West' Growth."

8. Van Oldenborgh et al., "Attribution of Extreme Rainfall From Hurricane Harvey."

9. Emanuel, "Assessing the Present and Future Probability of Hurricane Harvey's Rainfall"; Risser and Wehner, "Attributed Human-Induced Changes in Extreme Precipitation During Hurricane Harvey"; Wang et al., "Quantitative Attribution of Climate Effects on Hurricane Harvey's Extreme Rainfall."

10. Associated Press (in *Breitbart*), "Warming Made Harvey's Deluge 3 Times More Likely."

11. Achenbach, "Global Warming Boosted Hurricane Harvey's Rainfall by at Least 15 Percent."

12. Boburg and Reinhard, "Houston's 'Wild West' Growth."

13. Durkin, "North Carolina Didn't Like Science on Sea Levels."

14. Pilkey, "Sea-Level Rise Is Here."

15. Duncan, "Hurricane Florence Was Among the Costliest Disasters on Record."

16. Stott, Stone, and Allen, "Human Contribution to the European Heatwave of 2003."

17. Anderson et al., "The Dangers of Disaster-Driven Responses to Climate Change."

18. Schiermeier, "Droughts, Heatwaves and Floods."

19. King and Harrington, "The Inequality of Climate Change."

Chapter 7: Facts Not Fatalism: Identifying the Causes of Disasters in Order to Act

1. Deutsche Gesellschaft für Internationale Zusammenarbeit, *Climate Change Realities in Small Island Developing States in the Caribbean.*

2. Hallegatte, Bangalore, et al., *Shock Waves;* Hallegatte, Lecocq, and de Perthuis, "Designing Climate Change Adaption Policies"; Hallegatte, Vogt-Schlib, et al., *Unbreakable.*

3. United Nations Sustainable Development Goals, "Climate Action."

4. Oxfam, "A Climate in Crisis."

5. Adhikari, Nejadhashemi, and Woznicki, "Climate Change and Eastern Africa."

6. Wanzala, "Irrigation on Rise in Africa."

7. Uhe et al., "Attributing Drivers of the 2016 Kenyan Drought"; Philip et al., "Attribution Analysis of the Ethiopian Drought of 2015."

8. Eriksen and Marin, "Sustainable Adaptation under Adverse Development?"

9. Otto and van Aalst, "Droughts in East Africa."

10. Uhe et al., "Severe Drought in Kenya, 2016–17."

11. Chemweno, "Climate Scientists Warn of Worse Drought Situation Ahead."

12. BNP Paribas, "Climate Change: Stimulating Effective Adaptation Programs in Africa."

13. Oxfam, "A Climate in Crisis."

14. Klepp, "Climate Change and Migration."

15. Bedarff and Jakobeit, *Climate Change, Migration, and Displacement.*

16. Van Oldenborgh, van Urk, and Allen, "The Absence of a Role of Climate Change in the 2011 Thailand Floods."

17. Chari, "No Water, No Work."

Chapter 8: A Question of Justice: The Cost of Climate Change and the Responsibilities of Industrialized Countries

1. Cornwall, "As Sea Levels Rise, Bangladeshi Islanders Must Decide."

2. Philip et al., "Attributing the 2017 Bangladesh Floods."

3. Cornwall, "As Sea Levels Rise, Bangladeshi Islanders Must Decide."

4. Ward, *Under a Green Sky.*

5. Messenger, "A New Take on the World's Carbon Footprint (Graphic)."

6. Schiermeier, "Droughts, Heatwaves and Floods."

7. United Nations, *Paris Agreement.*

8. United Nations Framework Convention on Climate Change, *Approaches to Address Loss and Damage*, 3.

9. James et al., "Characterizing Loss and Damage."

10. Nobel Prize, "William D. Nordhaus."

11. Carbon Pricing Leadership Coalition, "More Than Eight-Fold Leap Over Four Years."

12. Frame et al., *Estimating Financial Costs of Climate Change in New Zealand.*

13. Energy and Climate Intelligence Unit, *Heavy Weather.*

Chapter 9: Countries and Corporations on Trial

1. Moloney, "Colombia's Top Court Orders Government to Protect Amazon Forest."

2. República de Colombia, Corte Suprema de Justicia.

3. Global Carbon Project, *Global Carbon Budget.*

4. Urgenda, "The Urgenda Climate Case Against the Dutch Government."

5. Farand, "Nine-Year-Old Girl Files Lawsuit Against Indian Government."

6. Gabbatiss, "EU Taken to Court by Families in 'People's Climate Case.'"

7. KlimaSeniorinnen, "English Summary of Our Climate Case."

8. Sabin Center for Climate Change Law, http://climatecasechart.com/about/.

9. Grantham Research Institute on Climate Change and the Environment, "Climate Change Laws of the World."

10. Thurau, "Germany's Angela Merkel No Longer Leading."

11. Staude, "Cañete gibt 45-Prozent-Ziel auf."

12. Bethge, "Drei Bauernfamilien verklagen die Bundesregierung."

13. United Nations Environment Programme, *The Status of Climate Change Litigation.*

14. Rauf, "Clean Break: Kennedy Supreme Court Exit."

15. United States District Court for the Northern District of California, Oakland Division, Native Village of Kivalina v. ExxonMobil Corporation et al.

16. Barringer, "Flooded Village Files Suit."

17. Nugent, "Climate Change Could Destroy This Peruvian Farmer's Home."

18. Nugent, "Climate Change Could Destroy This Peruvian Farmer's Home."

19. Nugent, "Climate Change Could Destroy This Peruvian Farmer's Home."

20. German Federal Ministry of Justice, *German Civil Code*.

21. Starr, "Just 90 Companies Are to Blame for Most Climate Change."

22. Ekwurzel et al., "The Rise in Global Atmospheric CO_2."

23. Olszynski, Mascher, and Doelle, "From Smokes to Smokestacks."

24. Marjanac, Patton, and Thornton, "Acts of God, Human Influence and Litigation"; Marjanac and Patton, "Extreme Weather Event Attribution Science and Climate Change Litigation"; McCormick et al., "Science in Litigation, the Third Branch of U.S. Climate Policy."

25. Marjanac and Patton, "Extreme Weather Event Attribution Science and Climate Change Litigation."

26. Skeie et al., "Perspective Has a Strong Effect."

27. Kasprak, "Did a 1912 Newspaper Article Predict Global Warming?"

28. Skeie et al., "Perspective Has a Strong Effect."

29. Skeie et al., "Perspective Has a Strong Effect."

30. Otto et al., "Assigning Historic Responsibility for Extreme Weather Events."

31. Frame et al., *Estimating Financial Costs of Climate Change in New Zealand*.

Chapter 10: Climate Change in Everyday Life: Seeing the Weather From a New Perspective

1. Domonoske, "Drought in Central Europe Reveals Cautionary 'Hunger Stones.'"

2. World Weather Attribution, "Heatwave in Northern Europe, Summer 2018."

3. Cockburn, "Fears Over Climate Change Hit Highest Level in a Decade."

4. Agence France-Presse, "'Unprecedented' Japan Heatwave Kills 65 People in a Week."

5. Schiermeier, "Droughts, Heatwaves and Floods."

6. Schaller et al., "Human Influence on Climate in the 2014 Southern England Winter Floods."

7. Climate Coalition, "Game Changer."

8. Otto et al., "Climate Change Increases the Probability of Heavy Rains."

9. Samenow, "Former Hurricane Ophelia Rocks Ireland With 100-Mph Wind Gusts."

Epilogue

1. Simon, "'Bad News' and 'Despair.'"

2. Otto et al., "Attributing High-Impact Extreme Events Across Time-scales."

BIBLIOGRAPHY

1. Multimedia

Competitive Enterprise Institute. "Global Warming—'Energy.'" YouTube video, May 18, 2006. https://www.youtube.com/watch?v=7sGKvDNdJNA.

EIKE (European Climate and Energy Institute). "Marc Morano: US-Klimapolitik nach 10 Monaten Trump - Stunde Null für Klimaalarmisten? (IKEK-11)." YouTube video in English, January 18, 2018. https://www.youtube.com/watch?v=38oxwzkouTs&t=607s.

2. Websites

Climate*prediction*.net. https://www.climateprediction.net/.

Old Weather. https://www.oldweather.org/.

Sabin Center for Climate Change Law. http://climatecasechart.com/about/.

3. Literature

Achenbach, Joel. "Global Warming Boosted Hurricane Harvey's Rainfall by at Least 15 Percent, Studies Find." *Washington Post*, December 13, 2017. https://www.washingtonpost.com/news/post-nation/wp/2017/12/13/global-warming-boosted-hurricane-harveys-rainfall-by-at-least-15-percent-studies-find/.

Adhikari, Umesh, A. Pouyan Nejadhashemi, and Sean A. Woznicki. "Climate Change and Eastern Africa: A Review of Impact on Major Crops." *Food and Energy Security* 4, no. 2 (2015): 110–132. https://onlinelibrary.wiley.com/doi/full/10.1002/fes3.61.

Agence France-Presse. "'Unprecedented' Japan Heatwave Kills 65 People in a Week." *The Guardian*, July 24, 2018. https://www.theguardian.com/world/2018/jul/24/unprecedented-japan-heatwave-kills-65-people-week.

Allen, Myles. "Liability for Climate Change." *Nature* 421, no. 6926 (2003): 891–892. https:/doi.org/10.1038/421891a.

Amadeo, Kimberly. "Hurricane Harvey Facts, Damage and Costs." *The Balance*, updated June 25, 2019. https://www.thebalance.com/hurricane-harvey-facts-damage-costs-4150087.

Amadeo, Kimberly. "Hurricane Irma Facts, Damage, and Costs." *The Balance*, updated October 21, 2019. https://www.thebalance.com/hurricane-irma-facts-timeline-damage-costs-4150395.

American Red Cross. "Hurricane Harvey—Red Cross on the Scene." August 30, 2017. https://www.redcross.org/about-us/news-and-events/news/Hurricane-Harvey-Red-Cross-on-the-Scene.html.

Anderson, Sarah E., Ryan R. Bart, Maureen C. Kennedy, Andrew J. MacDonald, Max A. Moritz, Andrew J. Plantinga, Christina L. Tague, and Matthew Wibbenmeyer. "The Dangers of Disaster-Driven Responses to Climate Change." *Nature Climate Change* 8 (2018): 651–653. https://doi.org/10.1038/s41558-018-0208-8.

Associated Press. "Studies: Warming Made Harvey's Deluge 3 Times More Likely." *Breitbart*, December 14, 2017. https://www.breitbart.com/news/studies-warming-made-harveys-deluge-3-times-more-likely/.

Barringer, Felicity. "Flooded Village Files Suit, Citing Corporate Link to Climate Change." *New York Times*, February 27, 2008. https://www.nytimes.com/2008/02/27/us/27alaska.html.

Bedarff, Hildegard, and Cord Jakobeit. *Climate Change, Migration, and Displacement: The Underestimated Disaster.* Greenpeace Germany, May 2017. https://www.greenpeace.de/sites/www.greenpeace.de/files/20170524-greenpeace-studie-climate-change-migration-displacement-engl.pdf.

Bethge, Philip. "Drei Bauernfamilien verklagen die Bundesregierung." *Der Spiegel*, October 26, 2018. http://www.spiegel.de/wissenschaft/natur/klima-klage-gegen-bundesregierung-a-1235300.html.

Blake, Eric S., and David A. Zelinsky. *National Hurricane Center Tropical Cyclone Report: Hurricane Harvey.* NOAA and National Weather Service, May 9, 2018. https://www.nhc.noaa.gov/data/tcr/AL092017_Harvey.pdf.

BNP Paribas. "Climate Change: Stimulating Effective Adaptation Programs in Africa." December 1, 2017. https://group.bnpparibas/en/news/climate-change-stimulating-effective-adaptation-programs-africa.

Boburg, Shawn, and Beth Reinhard. "Houston's 'Wild West' Growth." *Washington Post*, August 29, 2017. https://www.washingtonpost.com/graphics/2017/investigations/harvey-urban-planning/.

Box, George E. P., and Norman R. Draper. *Empirical Model-Building and Response Surfaces*. New York: John Wiley and Sons, 1987.

Boyd, Emily, Rachel A. James, Richard G. Jones, Hannah R. Young, and Friederike E. L. Otto. "A Typology of Loss and Damage Perspectives." *Nature Climate Change* 7 (2017): 723–729. https://doi.org/10.1038/nclimate3389.

Brulle, Robert J. "Institutionalizing Delay: Foundation Funding and the Creation of U.S. Climate Change Counter-Movement Organizations." *Climatic Change* 122, no. 4 (2014): 681–694. https://doi.org/10.1007/s10584-013-1018-7.

Bussewitz, Cathy. "New York Says Exxon Misled Investors About Climate Risks." Associated Press, October 24, 2018. https://apnews.com/1064ae274dff49eab1989a1323b84d81.

Carbon Brief. "Q&A: How Do Climate Models Work?" January 15, 2018. https://www.carbonbrief.org/qa-how-do-climate-models-work.

Carbon Pricing Leadership Coalition. "More Than Eight-Fold Leap Over Four Years in Global Companies Pricing Carbon Into Business Plans." October 13, 2017. https://www.carbonpricingleadership.org/news/cdp-report-2017.

Chappell, Bill. "National Weather Service Adds New Colors So It Can Map Harvey's Rains." NPR, August 28, 2017. https://www.npr.org/sections/thetwo-way/2017/08/28/546776542/national-weather-service-adds-new-colors-so-it-can-map-harveys-rains.

Chari, Mridula. "No Water, No Work: Why Drought Migrants in Mumbai Are Reluctant to Go Home." *Scroll.in*, May 31, 2016. https://scroll.in/article/809010/no-water-no-work-why-drought-migrants-in-mumbai-are-reluctant-to-go-home.

Chemweno, Brigid. "Climate Scientists Warn of Worse Drought Situation Ahead." *Standard Digital*, March 23, 2017. https://www.standardmedia.

co.ke/business/article/2001233757/climate-scientists-warn-of-worse-drought-situation-ahead.

Circuit Court for Baltimore City: Mayor and City Council of Baltimore, Plaintiff, v. BP PLC et al., Defendants. July 20, 2018. https://law.baltimorecity.gov/sites/default/files/Climate%20Change%20Complaint.pdf.

Climate Coalition. "Game Changer: How Climate Change Is Impacting Sports in the UK." Accessed October 1, 2019. https://static1.squarespace.com/static/58b40fe1be65940cc4889d33/t/5a85c91e9140b71180b a91e0/1518717218061/The+Climate+Coalition_Game+Changer.pdf.

Climatefiles. "1982 Exxon Presentation on 'CO$_2$ Greenhouse Effect' and Exxon Climate Modeling." Accessed October 1, 2019. http://www.climatefiles.com/exxonmobil/august-24-1982-exxon-presentation-on-co2-greenhouse-effect-and-exxon-climate-modeling/.

Cockburn, Harry. "Fears Over Climate Change Hit Highest Level in a Decade Following Heatwave, Study Says." *The Independent*, September 4, 2018. https://www.independent.co.uk/environment/climate-change-heatwave-global-warming-opinium-poll-leo-barasi-a8522901.html.

Collier, Kiah, and Neena Satija. "A Year Before Harvey, Houston-Area Flood Control Chief Saw No 'Looming Issues.'" *Texas Tribune*, September 7, 2017. https://www.texastribune.org/2017/09/07/conversation-former-harris-county-flood-control-chief/.

Cornwall, Warren. "As Sea Levels Rise, Bangladeshi Islanders Must Decide Between Keeping the Water Out—or Letting It In." *ScienceMag*, March 1, 2018. https://www.sciencemag.org/news/2018/03/sea-levels-rise-bangladeshi-islanders-must-decide-between-keeping-water-out-or-letting.

Coy, Peter, and Christopher Flavelle. "Harvey Wasn't Just Bad Weather. It Was Bad City Planning." *Bloomberg*, August 31, 2017. https://www.bloomberg.com/news/features/2017-08-31/a-hard-rain-and-a-hard-lesson-for-houston.

Davidson Sorkin, Amy. "What Has Hurricane Harvey Taught Donald Trump in Texas?" *New Yorker*, August 29, 2017. https://www.newyorker.com/news/daily-comment/what-did-donald-trump-learn-in-texas.

Deutsche Gesellschaft für Internationale Zusammenarbeit. *Climate Change Realities in Small Island Developing States in the Caribbean.* Commissioned by the Global Programme on Risk Assessment and Management for Adaptation to Climate Change, May 2017. https://www.adaptation community.net/wp-content/uploads/2017/05/Grenada-Study.pdf.

Di Liberto, Tom. "India Heat Wave Kills Thousands." Climate.gov, June 9, 2015. https://www.climate.gov/news-features/event-tracker/india-heat-wave-kills-thousands.

Dole, Randall, Martin Hoerling, Judith Perlwitz, Jon Eischeid, Philip Pegion, Tao Zhang, Xiao-Wei Quan, Taiyi Xu, and Donald Murray. "Was There a Basis for Anticipating the 2010 Russian Heat Wave?" *Geophysical Research Letters* 38, no. 6 (2011): L06702. https://doi.org/10.1029/2010GL046582.

Domonoske, Camila. "Drought in Central Europe Reveals Cautionary 'Hunger Stones' in Czech River." NPR, August 24, 2018. https://www.npr.org/2018/08/24/641331544/drought-in-central-europe-reveals-cautionary-hunger-stones-in-czech-river?t=1572278224829.

Duncan, Charles. "Hurricane Florence Was Among the Costliest Disasters on Record. Here's NOAA's Tally." *News and Observer*, February 8, 2019. https://www.newsobserver.com/news/state/north-carolina/article 225974185.html.

Durkin, Erin. "North Carolina Didn't Like Science on Sea Levels . . . So Passed a Law Against It." *The Guardian*, September 12, 2018. https://www.theguardian.com/us-news/2018/sep/12/north-carolina-didnt-like-science-on-sea-levels-so-passed-a-law-against-it.

Ekwurzel, B., J. Boneham, M. W. Dalton, R. Heede, R. J. Mera, M. R. Allen, and P. C. Frumhoff. "The Rise in Global Atmospheric CO_2, Surface Temperature, and Sea Level From Emissions Traced to Major Carbon Producers." *Climatic Change* 144, no. 4 (2017): 579–590. https://doi.org/10.1007/s10584-017-1978-0.

Ellen. "Fox News' *Outnumbered* Ignores Impact of Climate Change on Hurricane Harvey's Epic Intensity." *NewsHounds*, August 25, 2017. http://www.newshounds.us/fox_outnumbered_ignores_impact_of_climate_change_hurricane_harvey_intensity_082517.

Emanuel, Kerry. "Assessing the Present and Future Probability of Hurricane Harvey's Rainfall." *Proceedings of the National Academy of Sciences* 114, no. 48 (2017): 12681–12684. https://doi.org/10.1073/pnas.1716222114.

Energy and Climate Intelligence Unit. *Heavy Weather.* December 11, 2017. https://eciu.net/news-and-events/reports/2017/heavy-weather.

Eriksen, Siri H., and Andrei Marin. "Sustainable Adaptation Under Adverse Development? Lessons From Ethiopia." In *Climate Change Adaptation and Development: Transforming Paradigms and Practices,* edited by Tor Håkon Inderberg, Siri Eriksen, Karen O'Brien, and Linda Sygna, 178–199. Oxford: Routledge, 2015.

ExxonMobil. "ExxonMobil Allocates $500,000 for Gulf Coast Community Hurricane Relief Efforts." News release, August 25, 2017. http://news.exxonmobil.com/press-release/exxonmobil-allocates-500000-gulf-coast-community-hurricane-relief-efforts.

Farand, Chloe. "Nine-Year-Old Girl Files Lawsuit Against Indian Government Over Failure to Take Ambitious Climate Action." *The Independent,* April 1, 2017. https://www.independent.co.uk/environment/nine-ridhima-pandey-court-case-indian-government-climate-change-uttarakhand-a7661971.html.

Fischedick, Manfred, Klaus Görner, and Margit Thomeczek. *CO2: Abtrennung, Speicherung, Nutzung: Ganzheitliche Bewertung von Energiewirtschaft und Industrie.* Heidelberg: Springer, 2015.

Fountain, Henry. "Looking, Quickly, for the Fingerprints of Climate Change." *New York Times,* August 1, 2016. https://www.nytimes.com/2016/08/02/science/looking-quickly-for-the-fingerprints-of-climate-change.html.

Frame, Dave, Suzanne Rosier, Trevor Carey-Smith, Luke Harrington, Sam Dean, and Ilan Noy. *Estimating Financial Costs of Climate Change in New Zealand.* New Zealand Climate Research Institute and NIWA, April 21, 2018. https://treasury.govt.nz/sites/default/files/2018-08/LSF-estimating-financial-cost-of-climate-change-in-nz.pdf.

Franta, Benjamin. "Early Oil Industry Knowledge of CO_2 and Global Warming." *Nature Climate Change* 8 (2018): 1024–1025. https://doi.org/10.1038/s41558-018-0349-9.

Gabbatiss, Josh. "EU Taken to Court by Families in 'People's Climate Case' Over Inadequate 2030 Emissions Target." *The Independent,* May 24, 2018. https://www.independent.co.uk/news/world/europe/eu-emissions-targets-peoples-climate-case-change-european-parliament-global-warming-a8367146.html.

German Federal Ministry of Justice and Consumer Protection, and Federal Office of Justice. *German Civil Code*. English translation. Accessed October 2, 2019. http://www.gesetze-im-internet.de/englisch_bgb/englisch_bgb.pdf.

Gibbons, Brendan. "Harvey's Intensity and Rainfall Potential Tied to Global Warming." *San Antonio Express-News*, August 25, 2017. https://www.expressnews.com/news/local/article/Harvey-s-intensity-and-rainfall-potential-tied-11957010.php.

Gilmer, Bill. "Proximity Counts: How Houston Dominates the Oil Industry." *Forbes*, August 22, 2018. https://www.forbes.com/sites/uhenergy/2018/08/22/proximity-counts-how-houston-dominates-the-oil-industry/#d19e79610782.

Global Carbon Project. *Global Carbon Budget*. December 5, 2018. http://www.globalcarbonproject.org/carbonbudget/.

Grande, Bette, Alan Carlin, John Coleman, Frederick D. Palmer, Timothy Ball, S. T. Karnick, Paul Driessen, Daniel Sutter, William Briggs, Joseph S. D'Aleo, and Tom Harris. "Heartland Institute Experts Comment on Hurricane Harvey." News release, August 25, 2017. https://www.heartland.org/news-opinion/news/press-release-heartland-institute-experts-comment-on-hurricane-harvey.

Grantham Research Institute on Climate Change and the Environment. "Climate Change Laws of the World." Databases. Accessed October 2, 2019. http://www.lse.ac.uk/GranthamInstitute/climate-change-laws-of-the-world/.

Gustin, Georgina. "Judge Dismisses Youth Climate Change Lawsuit in Washington State." *Inside Climate News*, August 15, 2018. https://insideclimatenews.org/news/15082018/youth-climate-change-lawsuit-dismissed-washington-state-greenhouse-gas-emissions.

Hallegatte, Stephane, Mook Bangalore, Laura Bonzanigo, Marianne Fay, Tamaro Kane, Ulf Narloch, Julie Rozenberg, David Treguer, and Adrien Vogt-Schilb. *Shock Waves: Managing the Impacts of Climate Change on Poverty*. Washington, D.C.: World Bank, 2016. https://openknowledge.worldbank.org/handle/10986/22787.

Hallegatte, Stephane, Franck Lecocq, and Christian de Perthuis. "Designing Climate Change Adaption Policies: An Economic Framework."

Policy Research working paper no. WPS 5568. World Bank, 2011. https://openknowledge.worldbank.org/handle/10986/3335.

Hallegatte, Stephane, Adrien Vogt-Schilb, Mook Bangalore, and Julie Rozenberg. *Unbreakable: Building the Resilience of the Poor in the Face of Natural Disasters.* Washington, D.C.: World Bank, 2017. https://openknowledge.worldbank.org/handle/10986/25335.

Hasemyer, David. "2 City Lawsuits Against Big Oil Dismissed, but That's Not the End of It." *Inside Climate News*, June 26, 2018. https://insideclimatenews.org/news/26062018/california-cities-climate-change-lawsuits-dismissed-fossil-fuels-industry-rising-sea-levels.

Hauser, Christine. "Hurricane Harvey Strengthens and Heads for Texas." *New York Times*, August 24, 2017. https://www.nytimes.com/2017/08/24/us/harvey-storm-hurricane-texas.html.

International Energy Agency. *World Energy Outlook 2012: Executive Summary.* http://www.iea.org/publications/freepublications/publication/English.pdf.

Jacques, Peter J., Riley E. Dunlap, and Mark Freeman. "The Organisation of Denial: Conservative Think Tanks and Environmental Scepticism." *Environmental Politics* 17, no. 3 (2008): 349–385. https://doi.org/10.1080/09644010802055576.

James, Rachel, Friederike Otto, Hannah Parker, Emily Boyd, Rosalind Cornforth, Daniel Mitchell, and Myles Allen. "Characterizing Loss and Damage From Climate Change." *Nature Climate Change* 4 (2015): 938–939. https://doi.org/10.1038/nclimate2411.

Kalhoefer, Kevin. "So Far, ABC and NBC Are Failing to Note the Link Between Harvey and Climate Change." Media Matters for America, August 31, 2017. https://www.mediamatters.org/nbc/so-far-abc-and-nbc-are-failing-note-link-between-harvey-and-climate-change.

Kasprak, Alex. "Did a 1912 Newspaper Article Predict Global Warming?" Snopes, October 18, 2016. https://www.snopes.com/fact-check/1912-article-global-warming/.

Keim, Brandon. "Russian Heat Wave Statistically Linked To Climate Change." *Wired*, October 24, 2011. https://www.wired.com/2011/10/russian-heat-climate-change/.

Kew, Sarah F., Sjoukje Y. Philip, Geert Jan van Oldenborgh, Friederike E. L. Otto, Robert Vautard, and Gerard van der Schrier. "The Exceptional Summer Heat Wave in Southern Europe 2017." *Bulletin of the American Meteorological Society* 100, no. 1, Special Suppl. (2019): s49–s53. https://doi.org/10.1175/BAMS-D-18-0109.1.

King, Andrew D., and Luke J. Harrington. "The Inequality of Climate Change from 1.5 to 2°C of Global Warming." *Geophysical Research Letters* 45, no. 10 (2018): 5030–5033. https://doi.org/10.1029/2018GL078430.

Klepp, Silja. "Climate Change and Migration." *Climate Science,* April 2017. https://doi.org/10.1093/acrefore/9780190228620.013.42.

Klimaretter.Info. "Rechter Unions-Flügel folgt Trump." June 4, 2017. http://www.klimaretter.info/politik/nachricht/23221-rechterunions-fluegel-folgt-trumps-klimakurs.

KlimaSeniorinnen. "English Summary of Our Climate Case." Accessed September 23, 2019. https://klimaseniorinnen.ch/english/.

Krugman, Paul. "Enemy of the Planet." *New York Times,* April 17, 2006. https://www.nytimes.com/2006/04/17/opinion/enemy-of-the-planet.html.

Kusnetz, Nicholas, and David Hasemyer. "Judge Dismisses New York City Climate Lawsuit Against 5 Oil Giants." *Inside Climate News,* July 19, 2018. https://insideclimatenews.org/news/19072018/judge-dismisses-nyc-climate-change-lawsuit-oil-industry-global-warming-adaptation-costs.

Lascaris, Dimitri. "Are Irma-Like Super Storms the 'New Normal'?" *Real News Network,* September 7, 2017. https://therealnews.com/stories/mmanno906report.

Lewandowsky, Stephan, and Klaus Oberauer. "Motivated Rejection of Science." *Current Directions in Psychological Science* 25, no. 4 (2016): 217–222. https://doi.org/10.1177/0963721416654436.

Littlemore, Richard. "Heartland Insider Exposes Institute's Budget and Strategy." *DeSmogBlog,* February 14, 2012. https://www.desmogblog.com/heartland-insider-exposes-institute-s-budget-and-strategy.

Lusk, Greg. "The Social Utility of Event Attribution: Liability, Adaptation, and Justice-Based Loss and Damage." *Climatic Change* 143, no. 1–2 (2017): 201–212. https://doi.org/10.1007/s10584-017-1967-3.

Mann, Michael E. "It's a Fact: Climate Change Made Hurricane Harvey More Deadly." *The Guardian*, August 28, 2017. https://www. theguardian.com/commentisfree/2017/aug/28/climate-change-hurricane-harvey-more-deadly.

Marjanac, Sophie, and Lindene Patton. "Extreme Weather Event Attribution Science and Climate Change Litigation: An Essential Step in the Causal Chain?" *Journal of Energy and Natural Resources Law* 36, no. 3 (2018): 265–298. https://doi.org/10.1080/02646811.2018.1451020.

Marjanac, Sophie, Lindene Patton, and James Thornton. "Acts of God, Human Influence and Litigation." *Nature Geoscience* 10 (2017): 616–619. https://doi.org/10.1038/ngeo3019.

Massey, N., T. Aina, C. Rye, F. E. L. Otto, S. Wilson, R. G. Jones, and M. R. Allen. "Have the Odds of Warm November Temperatures and of Cold December Temperatures in Central England Changed?" In "Explaining Extreme Events of 2011 From a Climate Perspective," edited by Thomas C. Peterson, Peter A. Stott, and Stephanie Herring. *Bulletin of the American Meteorological Society* 93, no. 7, Special Suppl. (2012): 1057–1059. https://doi.org/10.1175/BAMS-D-12-00021.1.

McCormick, Sabrina, Samuel J. Simmens, Robert L. Glicksman, LeRoy Paddock, Daniel Kim, Brittany Whited, and William Davies. "Science in Litigation, the Third Branch of U.S. Climate Policy." *Science* 357, no. 6355 (2017): 979–980. https://doi.org/10.1126/science.aao0412.

Media Research Center. "Has CNN Warped Meteorologist Chad Myers' View on Climate Change?" Accessed September 30, 2019. https://www. mrc.org/articles/has-cnn-warped-meteorologist-chad-myers-view-climate-change.

Messenger, Stephen. "A New Take on the World's Carbon Footprint (Graphic)." *Treehugger*, February 2, 2011. Graphic first appeared in *Miller-McCune Magazine*, created by Stanford Kay Studios, 2010. https://www.treehugger.com/corporate-responsibility/a-new-take-on-the-worlds-carbon-footprint-graphic.html.

Moloney, Anastasia. "Colombia's Top Court Orders Government to Protect Amazon Forest in Landmark Case." Reuters, April 6, 2018. https://www.reuters.com/article/us-colombia-deforestation-amazon/colombias-top-court-orders-government-to-protect-amazon-forest-in-landmark-case-idUSKCN1HD21Y.

National Academies of Sciences, Engineering, and Medicine. *Attribution of Extreme Weather Events in the Context of Climate Change.* Washington, D.C.: The National Academies Press, 2016. https://doi.org/10.17226/21852.

Nobel Prize, The. "William D. Nordhaus—Facts—2018." Accessed November 7, 2019. https://www.nobelprize.org/prizes/economic-sciences/2018/nordhaus/facts/.

Nugent, Ciara. "Climate Change Could Destroy This Peruvian Farmer's Home. Now He's Suing a European Energy Company for Damages." *Time*, October 5, 2018. http://time.com/5415225/RWE-lliuya-climate-change/.

Olszynski, Martin, Sharon Mascher, and Meinhard Doelle. "From Smokes to Smokestacks: Lessons from Tobacco for the Future of Climate Change Liability." *Georgetown Environmental Law Review*, April 24, 2017. https://papers.ssrn.com/sol3/papers.cfm?abstract_id=2957921.

Oreskes, Naomi, and Erik M. Conway. *Merchants of Doubt: How a Handful of Scientists Obscured the Truth on Issues From Tobacco Smoke to Global Warming.* London: Bloomsbury Press, 2010.

Otto, Friederike. "Link Between Hurricane Harvey and Climate Change Is Unclear." *Climate Change News*, August 28, 2017. https://www.climatechangenews.com/2017/08/28/link-hurricane-harvey-climate-change-unclear/.

Otto, Friederike, and Maarten van Aalst. "Droughts in East Africa: Some Headway in Unpacking What's Causing Them." *The Conversation*, July 11, 2017. https://theconversation.com/droughts-in-east-africa-some-headway-in-unpacking-whats-causing-them-75476.

Otto, Friederike E. L., Caio A. S. Coelho, Andrew King, Erin Coughlan de Perez, Yoshihide Wada, Geert Jan van Oldenborgh, Rein Haarsma, Karsten Haustein, Peter Uhe, Maarten van Aalst, José Antonio Aravéquia, Waldenio Almeida, and Heidi Cullen. "Factors Other Than Climate Change, Main Drivers of 2014/15 Water Shortage in Southeast Brazil." In "Explaining Extreme Events of 2014 from a Climate Perspective," edited by Stephanie C. Herring, Martin P. Hoerling, James P. Kossin, Thomas C. Peterson, and Peter A. Stott. *Bulletin of the American Meteorological Society* 96, no. 12, Special Suppl. (2015): S35–S40. https://doi.org/10.1175/BAMS-EEE_2014_ch8.1.

Otto, Friederike E. L., Sjoukje Philip, Sarah Kew, Sihan Li, Andrew King, and Heidi Cullen. "Attributing High-Impact Extreme Events Across Timescales—a Case Study of Four Different Types of Events." *Climatic Change* 149, no. 3–4 (2018): 399–412. https://doi.org/10.1007/s10584-018-2258-3.

Otto, Friederike E. L., Ragnhild B. Skeie, Jan S. Fuglestvedt, Terje Berntsen, and Myles R. Allen. "Assigning Historic Responsibility for Extreme Weather Events." *Nature Climate Change* 7 (2017): 757–759. https://doi.org/10.1038/nclimate3419.

Otto, Friederike E. L., Karin van der Wiel, Geert Jan van Oldenborgh, Sjoukje Philip, Sarah F. Kew, Peter Uhe, and Heidi Cullen. "Climate Change Increases the Probability of Heavy Rains in Northern England/Southern Scotland Like Those of Storm Desmond—a Real-Time Event Attribution Revisited." *Environmental Research Letters* 13, no. 2 (2017): 024006. https://doi.org/10.1088/1748-9326/aa9663.

Oxfam. "A Climate in Crisis: How Climate Change Is Making Drought and Humanitarian Disaster Worse in East Africa." Media briefing, April 27, 2017. https://oi-files-d8-prod.s3.eu-west-2.amazonaws.com/s3fs-public/mb-climate-crisis-east-africa-drought-270417-en.pdf.

Philip, Sjoukje, Sarah F. Kew, Geert Jan van Oldenborgh, Friederike Otto, Sarah O'Keefe, Karsten Haustein, Andrew King, Abiy Zegeye, Zewdu Eshetu, Kinfe Hailemariam, Roop Singh, Eddie Jjemba, Chris Funk, and Heidi Cullen. "Attribution Analysis of the Ethiopian Drought of 2015." *Journal of Climate* 31, no. 6 (2018): 2465–2486. https://doi.org/10.1175/JCLI-D-17-0274.1.

Philip, Sjoukje, Sarah Sparrow, Sarah F. Kew, Karin van der Wiel, Niko Wanders, Roop Singh, Ahmadul Hassan, Khaled Mohammed, Hammad Javid, Karsten Haustein, Friederike E. L. Otto, Feyera Hirpa, Ruksana H. Rimi, A. K. M. Saiful Islam, David C. H. Wallom, and Geert Jan van Oldenborgh. "Attributing the 2017 Bangladesh Floods From Meteorological and Hydrological Perspectives." *Hydrology and Earth System Sciences* 23, no. 3 (2019): 1409–1429. https://doi.org/10.5194/hess-2018-379.

Pierre-Louis, Kendra. "Climate Change to Blame for Deaths in 2003 Heat Wave." *Inside Climate News*, July 8, 2016. https://insideclimatenews.org/

news/07072016/climate-change-blame-deadliness-2003-heat-wave-new-study-paris-london.

Pilkey, Orrin H. "Sea-Level Rise Is Here. North Carolina Needs to Act." *News and Observer*, September 7, 2018. https://www.newsobserver.com/latest-news/article217954910.html.

Rauf, David S. "Clean Break: Kennedy Supreme Court Exit Could Upend Environmental Safeguards." *Scientific American*, July 3, 2018. https://www.scientificamerican.com/article/clean-break-kennedy-supreme-court-exit-could-upend-environmental-safeguards/.

República de Colombia, Corte Suprema de Justicia. STC4360-2018, Radicación n.° 11001-22-03-000-2018-00319-01. 2018. https://www.dejusticia.org/wp-content/uploads/2018/01/Fallo-Corte-Suprema-de-Justicia-Litigio-Cambio-Clim%C3%A1tico.pdf?x54537.

Rice, Doyle. "Harvey to Be Costliest Natural Disaster in U.S. History, Estimated Cost of $190 Billion." *USA Today*, August 31, 2017. https://www.usatoday.com/story/weather/2017/08/30/harvey-costliest-natural-disaster-u-s-history-estimated-cost-160-billion/615708001/.

Risser, Mark D., and Michael F. Wehner. "Attributable Human-Induced Changes in the Likelihood and Magnitude of the Observed Extreme Precipitation During Hurricane Harvey." *Geophysical Research Letters* 44, no. 24 (2017): 12457-12464. https://doi.org/10.1002/2017GL075888.

Robine, J. M., S. L. Cheung, S. Le Roy, H. van Oyen, and F. R. Herrmann. *Report on Excess Mortality in Europe During Summer 2003*. E.U. Community Action Programme for Public Health, February 28, 2007. http://ec.europa.eu/health/ph_projects/2005/action1/docs/action1_2005_a2_15_en.pdf.

Samenow, Jason. "60 Inches of Rain Fell From Hurricane Harvey in Texas, Shattering U.S. Storm Record." *Washington Post*, September 22, 2017. https://www.washingtonpost.com/news/capital-weather-gang/wp/2017/08/29/harvey-marks-the-most-extreme-rain-event-in-u-s-history/.

Samenow, Jason. "Former Hurricane Ophelia Rocks Ireland With 100-Mph Wind Gusts." *Washington Post*, October 16, 2017. https://www.washingtonpost.com/news/capital-weather-gang/wp/2017/10/16/former-hurricane-ophelia-rocks-ireland-with-100-mph-wind-gusts/.

Schaller, Nathalie, Alison L. Kay, Rob Lamb, Neil R. Massey, Geert Jan van Oldenborgh, Friederike E. L. Otto, Sarah N. Sparrow, Robert Vautard,

Pascal Yiou, Ian Ashpole, Andy Bowery, Susan M. Crooks, Karsten Haustein, Chris Huntingford, William J. Ingram, Richard G. Jones, Tim Legg, Jonathan Miller, Jessica Skeggs, David Wallom, Antje Weisheimer, Simon Wilson, Peter A. Stott, and Myles R. Allen. "Human Influence on Climate in the 2014 Southern England Winter Floods and Their Impacts." *Nature Climate Change* 6 (2016): 627–634. https://doi.org/10.1038/nclimate2927.

Schaller, Nathalie, Friederike E. L. Otto, Geert Jan van Oldenborgh, Neil R. Massey, Sarah Sparrow, and Myles R. Allen. "The Heavy Precipitation Event of May–June 2013 in the Upper Danube and Elbe Basins." In "Explaining Extreme Events of 2013 From a Climate Perspective," edited by Stephanie C. Herring, Martin P. Hoerling, Thomas C. Peterson, and Peter A. Stott. *Bulletin of the American Meteorological Society* 95, no. 9, Special Suppl. (2014): S69–S72. https://www2.ametsoc.org/ams/assets/File/publications/BAMS_EEE_2013_Full_Report.pdf.

Schiermeier, Quirin. "Droughts, Heatwaves and Floods: How to Tell When Climate Change Is to Blame." *Nature* news feature, July 30, 2018. https://www.nature.com/articles/d41586-018-05849-9.

School of Geography and the Environment, University of Oxford. "About the Radcliffe Meteorological Station's Records." Last modified December 1, 2016. http://www.geog.ox.ac.uk/research/climate/rms/about.html.

Schwartz, John. "New York Sues Exxon Mobil, Saying It Deceived Shareholders on Climate Change." *New York Times*, October 24, 2018. https://www.nytimes.com/2018/10/24/climate/exxon-lawsuit-climate-change.html.

Simon, Frédéric. "'Bad News' and 'Despair': Global Carbon Emissions to Hit New Record in 2018, IEA Says." *Euractiv*, October 18, 2018. https://www.euractiv.com/section/climate-environment/news/bad-news-and-despair-global-carbon-emissions-to-hit-new-record-in-2018-iea-says/.

Sims, Shannon. "The U.S. Flooded One of Houston's Richest Neighborhoods to Save Everyone Else." *Bloomberg Businessweek*, November 16, 2017. https://www.bloomberg.com/news/features/2017-11-16/the-u-s-flooded-one-of-houston-s-richest-neighborhoods-to-save-everyone-else.

Sippel, Sebastian, and Friederike E. L. Otto. "Beyond Climatological Extremes—Assessing How the Odds of Hydrometeorological Extreme Events in South-East Europe Change in a Warming Climate." *Climate*

Change 125, no. 3–4 (2014): 381–398. https://doi.org/10.1007/s10584-014-1153-9.

Skeie, Ragnhild B., Jan Fuglestvedt, Terje Berntsen, Glen P. Peters, Robbie Andrew, Myles Allen, and Steffen Kallbekken. "Perspective Has a Strong Effect on the Calculation of Historical Contributions to Global Warming." *Environmental Research Letters* 12, no. 2 (2017): 024022. https://doi.org/10.1088/1748-9326/aa5b0a.

Smith-Spark, Laura. "Hambach Forest Clearance Halted by German Court." CNN World, October 5, 2018. https://edition.cnn.com/2018/10/05/europe/germany-hambach-forest-court-intl/index.html.

Starr, Douglas. "Just 90 Companies Are to Blame for Most Climate Change, This 'Carbon Accountant' Says." *ScienceMag*, August 25, 2016. https://www.sciencemag.org/news/2016/08/just-90-companies-are-to-blame-most-climate-change-carbon-accountant-says.

Staude, Jörg. "Cañete gibt 45-Prozent-Ziel auf." *klimareporter°*, September 28, 2018. https://www.klimareporter.de/europaische-union/canete-gibt-45-prozent-ziel-auf.

Stern, Nicholas. *The Economics of Climate Change: The Stern Review.* Cambridge: Cambridge University Press, 2007. https://doi.org/10.1017/CBO9780511817434.

Stott, Peter A., D. A. Stone, and M. R. Allen. "Human Contribution to the European Heatwave of 2003." *Nature* 432, no. 7017 (2004): 610–614. https://doi.org/10.1038/nature03089.

Subramanian, Meera. "In Georgia's Peach Orchards, Warm Winters Raise Specter of Climate Change." *Inside Climate News*, August 31, 2017. https://insideclimatenews.org/news/31082017/climate-change-georgia-peach-harvest-warm-weather-crop-risk-farmers.

Supran, Geoffrey, and Naomi Oreskes. "Assessing ExxonMobil's Climate Change Communications (1977–2014)." *Environmental Research Letters* 12, no. 8 (2017): 084019. http://iopscience.iop.org/article/10.1088/1748-9326/aa815f.

Supreme Court of the State of New York, County of New York. People of the State of New York, by Barbara D. Underwood, Attorney General of the State of New York, Plaintiff, v. Exxon Mobil Corporation, Defendant. Summons, October 24, 2018. https://ag.ny.gov/sites/default/files/summons_and_complaint_0.pdf.

Sweney, Mark. "BBC Radio 4 Broke Accuracy Rules in Nigel Lawson Climate Change Interview." *The Guardian*, April 9, 2018. https://www.theguardian.com/environment/2018/apr/09/bbc-radio-4-broke-impartiality-rules-in-nigel-lawson-climate-change-interview.

Thurau, Jens. "Germany's Angela Merkel No Longer Leading the Charge on Climate Change." DW, October 8, 2018. https://www.dw.com/en/germanys-angela-merkel-no-longer-leading-the-charge-on-climate-change/a-45803875.

Timbal, Bertrand, Julie M. Arblaster, and Scott Power. "Attribution of the Late-Twentieth-Century Rainfall Decline in Southwest Australia." *Journal of Climate* 19 (2006): 2046–2062. https://doi.org/10.1175/JCLI3817.1.

Uhe, Peter, Sjoukje Philip, Sarah Kew, Kasturi Shah, Joyce Kimutai, Emmah Mwangi, Geert Jan van Oldenborgh, Roop Singh, Julie Arrighi, Eddie Jjemba, Heidi Cullen, and Friederike Otto. "Attributing Drivers of the 2016 Kenyan Drought." *International Journal of Climatology* 38, no. S1 (2018): e554–e568. https://doi.org/10.1002/joc.5389.

Uhe, Peter, Sjoukje Philip, Sarah Kew, Kasturi Shah, Joyce Kimutai, Friederike Otto, Geert Jan van Oldenborgh, Roop Singh, Julie Arrighi, and Heidi Cullen. "Severe Drought in Kenya, 2016–17." World Weather Attribution, March 23, 2017. https://www.worldweatherattribution.org/kenya-drought-2016/.

Union of Concerned Scientists. *Smoke, Mirrors, and Hot Air: How Exxon-Mobil Uses Big Tobacco's Tactics to Manufacture Uncertainty on Climate Science.* January 2007. https://www.ucsusa.org/sites/default/files/legacy/assets/documents/global_warming/exxon_report.pdf.

United Nations. *Paris Agreement.* 2015. https://unfccc.int/sites/default/files/english_paris_agreement.pdf.

United Nations Environment Programme. *The Status of Climate Change Litigation: A Global Review.* May 2017. https://wedocs.unep.org/handle/20.500.11822/20767.

United Nations Framework Convention on Climate Change. *Adoption of the Paris Agreement.* December 12, 2015. https://unfccc.int/resource/docs/2015/cop21/eng/l09.pdf.

United Nations Framework Convention on Climate Change. *Approaches to Address Loss and Damage Associated With Climate Change Impacts in Developing Countries That Are Particularly Vulnerable to the Adverse Effects of Climate Change to Enhance Adaptive Capacity.* November 15, 2012. https://unfccc.int/resource/docs/2012/sbi/eng/inf14.pdf.

United Nations Sustainable Development Goals. "Climate Action." Accessed October 2, 2019. https://www.un.org/sustainabledevelopment/climate-action/.

United States District Court for the Northern District of California, Oakland Division. Native Village of Kivalina, and City of Kivalina, Plaintiffs, v. ExxonMobil Corporation et al., Defendants. Case No. C08-1138 SBA. Order granting defendants' motions to dismiss for lack of subject matter jurisdiction, September 30, 2009. http://www.shopfloor.org/wp-content/uploads/kivalina-order-granting-motions-to-dismiss.pdf.

Urgenda. "The Urgenda Climate Case Against the Dutch Government." October 9, 2018. http://www.urgenda.nl/en/themas/climate-case/.

van der Wiel, Karin, Sarah B. Kapnick, Geert Jan van Oldenborgh, Kirien Whan, Sjoukje Philip, Gabriel A. Vecchi, Roop K. Singh, Julie Arrighi, and Heidi Cullen. "Rapid Attribution of the August 2016 Flood-Inducing Extreme Precipitation in South Louisiana to Climate Change." *Hydrology and Earth System Sciences* 21 (2017): 897–921. https://doi.org/10.5194/hess-21-897-2017.

van Oldenborgh, Geert Jan, Sjoukje Philip, Sarah Kew, Michiel van Weele, Peter Uhe, Friederike Otto, Roop Singh, Indrani Pai, Heidi Cullen, and Krishna AchutaRao. "Extreme Heat in India and Anthropogenic Climate Change." *Natural Hazards and Earth System Sciences* 18, no. 1 (2018): 365–381. https://doi.org/10.5194/nhess-18-365-2018.

van Oldenborgh, Geert Jan, Karin van der Wiel, Antonia Sebastian, Roop Singh, Julie Arrighi, Friederike Otto, Karsten Haustein, Sihan Li, Gabriel Vecchi, and Heidi Cullen. "Attribution of Extreme Rainfall From Hurricane Harvey, August 2017." *Environmental Research Letters* 12, no. 12 (2017): 124009. https://doi.org/10.1088/1748-9326/aa9ef2.

van Oldenborgh, Geert Jan, Anne van Urk, and Myles Allen. "The Absence of a Role of Climate Change in the 2011 Thailand Floods." In "Explaining Extreme Events of 2011 from a Climate Perspective," edited by Thomas C. Peterson, Peter A. Stott, and Stephanie Herring. *Bulletin of the American Meteorological Society* 93, no. 7 (2012): 1041–1067. https://doi.org/10.1175/BAMS-D-12-00021.1.

Vautard, Robert, Olivier Boucher, Geert Jan van Oldenborgh, Friederike Otto, Karsten Haustein, Martha M. Vogel, Sonia I. Seneviratne, Jean-Michel Soubeyroux, Michel Schneider, Agathe Drouin, Aurélien Ribes, Frank Kreienkamp, Peter Stott, and Maarten van Aalst. *Human Contribution to the Record-Breaking July 2019 Heat Wave in Western Europe*. World Weather Attribution, undated report. https://www.worldweatherattribution.org/wp-content/uploads/July2019heatwave.pdf.

Vidal, John. "Revealed: How Oil Giant Influenced Bush." *The Guardian*, June 8, 2005. https://www.theguardian.com/news/2005/jun/08/usnews.climatechange.

Wallace, Tim, Derek Watkins, Haeyoun Park, Anjali Singhvi, and Josh Williams. "How One Houston Suburb Ended Up in a Reservoir." *New York Times*, March 22, 2018. https://www.nytimes.com/interactive/2018/03/22/us/houston-harvey-flooding-reservoir.html.

Wang, S-Y Simon, Lin Zhao, Jin-Ho Yoon, Phil Klotzbach, and Robert R. Gillies. "Quantitative Attribution of Climate Effects on Hurricane Harvey's Extreme Rainfall in Texas." *Environmental Research Letters* 13, no. 5 (2018): 054014. https://doi.org/10.1088/1748-9326/aabb85.

Wanzala, Justus. "Irrigation on Rise in Africa as Farmers Face Erratic Weather." Reuters, September 9, 2016. https://www.reuters.com/article/us-africa-irrigation-farming/irrigation-on-rise-in-africa-as-farmers-face-erratic-weather-idUSKCN11F2DT.

Ward, Peter D. *Under a Green Sky: Global Warming, the Mass Extinctions of the Past, and What They Can Tell Us About Our Future*. New York: Harper Perennial, 2008.

World Weather Attribution. "Heatwave in Northern Europe, Summer 2018." July 28, 2018. https://www.worldweatherattribution.org/attribution-of-the-2018-heat-in-northern-europe/.

INDEX

169–70; changing climate change stance, 41–42, 44; lawsuits against, 171–74. *See also* ExxonMobil

energy usage, 42

England. *See* Great Britain

Environmental Change Institute (Oxford), 11, 176

Ethiopia, 132–36

Europe (European Union): 2003 heat wave, 54, 95–96, 122–23, 192; 2006 heat wave, 96, 123; 2015 heat wave, 63–65; 2017 heat wave, 28, 96; adaptation to extreme heat, 123; calculation of emissions responsibility, 176, 177; carbon price, 153; lawsuit against, 163; permanent attribution service, 125–26; responsibility for Argentinian heat wave, 178

Europe, 2018 heat wave: aridity and, 184n; attribution study, 109, 184–85; defining normal in relation to, 188–90; destruction from, 2; media response to study, 186–88; as new normal, 184–85, 187; public awareness about climate change and, 2, 182–84

European Centre for Medium-Range Weather Forecasts (ECMWF), 90

event attribution science: calculation of responsibility for specific extreme events, 178–79; for climate risk insurance,

159; definition, 4; in German Meteorological Service, 126–27; importance of, xiii–xv, 143, 193, 202–3; for lawsuits, 170, 174–76; limitations of, 103–5, 156; for loss and damage, 149, 154–55; need for global attribution services, 127–28; need for more studies, 190–91; official recognition of, 61–62, 66–67; permanent European service, 125–26; simulation of weather without climate change, 72–74; value for governments, 136–38; weather data for, 74–75. *See also* climate models; World Weather Attribution

event reconstruction, 26–28

extreme weather. *See* weather

ExxonMobil, 34–37, 41–42, 50, 51, 166, 170, 171

Finland, 184, 188–89

floods: Brahmaputra delta (2017), 147; climate change and, 192; event reconstruction and, 26–27; Germany (2013), 25, 103; Great Britain (2014 and 2017), 24–25, 81, 192; Japan (2018), 2; soccer and, 193–94; South China (2015), 154; Thailand (2010 and 2011), 142, 201. *See also* rainfall

Foote, Eunice Newton, x

fossil fuels. *See* energy corporations

Frame, David, 179

For more information about becoming a citizen scientist, getting involved in the world's largest climate model simulation, and supporting the work of the World Weather Attribution team, please visit climate*prediction*.net.

DAVID
SUZUKI
INSTITUTE

The David Suzuki Institute is a non-profit organization founded in 2010 to stimulate debate and action on environmental issues. The Institute and the David Suzuki Foundation both work to advance awareness of environmental issues important to all Canadians.

We invite you to support the activities of the Institute. For more information please contact us at:

David Suzuki Institute
219–2211 West 4th Avenue
Vancouver, BC v6k 4s2
info@davidsuzukiinstitute.org
604-742-2899
www.davidsuzukiinstitute.org

Cheques can be made payable to The David Suzuki Institute.